Public Investments in Energy Technology

To Our Loved Ones, Robin, Carol, and Scott

Public Investments in Energy Technology

Michael P. Gallaher

Director, Environmental, Technology, and Energy Economics Program, RTI International, USA

Albert N. Link

Professor of Economics, University of North Carolina at Greensboro, USA

Alan O'Connor

Senior Economist, Environmental, Technology, and Energy Economics Program, RTI International, USA

Edward Elgar

Cheltenham, UK • Northampton, MA, USA

Published by
Edward Elgar Publishing Limited
The Lypiatts
15 Lansdown Road
Cheltenham
Glos GL50 2JA
UK

Edward Elgar Publishing, Inc.
William Pratt House
9 Dewey Court
Northampton
Massachusetts 01060
USA

A catalogue record for this book
is available from the British Library

Library of Congress Control Number: 2011932900

ISBN 978 0 85793 157 3

Typeset by Servis Filmsetting Ltd, Stockport, Cheshire
Printed and bound by MPG Books Group, UK

Contents

Abbreviations

AEC Atomic Energy Commission
BCR benefit-cost ratio
BOS balance of systems
BTE brake thermal efficiency
CDC Centers for Disease Control and Prevention
CdTe cadmium telluride
CIS copper indium diselenide
COBRA Co-Benefits Risk Assessment model
COI cost of illness
CRF Combustion Research Facility
c-Si crystalline silicon
DOE Department of Energy
EERE The Office of Energy Efficiency and Renewable Energy
EFG edge-defined film-fed growth method
EIA Energy Information Administration
EPA Environmental Protection Agency
ERDA Energy Research and Development Administration
EOP Executive Office of the President
EVA ethylene vinyl acetate
FACA Federal Advisory Committee Act
FSA Flat-Plate Solar Array Project
GDP gross domestic product
GTP Geothermal Technologies Program
GHG greenhouse gas
INL Idaho National Laboratory
IRR internal rate of return
JPL Jet Propulsion Laboratory
LANL Los Alamos National Laboratory
LAS laser absorption spectrometry
LBNL Lawrence Berkeley National Laboratory
LCOE levelized cost of electricity
LDV laser Doppler velocimetry
LIF laser-induced fluorescence
LLNL Lawrence Livermore National Laboratory

LRS	laser Raman spectroscopy
MIT	Massachusetts Institute of Technology
MRAD	minor restricted-activity days
MW	megawatts
NASA	National Aeronautics and Space Administration
NIST	National Institute of Standards and Technology
NPV	net present value
NREL	National Renewable Energy Laboratory
NSF	National Science Foundation
OMB	Office of Management and Budget
OPEC	Organization of the Petroleum Exporting Countries
ORNL	Oak Ridge National Laboratory
PDC	polycrystalline diamond compact
PNGV	Partnership for a New Generation of Vehicles
PIV	particle image velocimetry
PV	photovoltaics
PVMaT	Photovoltaic Manufacturing Technology
R&D	research and development
RD&D	research, development and demonstration
SETP	Solar Energy Technology Program
SNL	Sandia National Laboratories
TFP	Thin-Film PV Partnership
UCC	Union Carbide Corporation
UNECE	United Nations Economic Commission for Europe
USGS	US Geological Survey
VSL	value of statistical life
VTP	Vehicle Technologies Program

About the authors

Michael P. Gallaher is the director of RTI's Environmental, Technology, and Energy Economics Program and has more than 15 years of experience leading projects for the US Department of Energy (DOE), National Institute of Standards and Technology (NIST), the US Environmental Protection Agency (EPA) and other federal agencies modeling the economic impact of new technologies. Dr. Gallaher specializes in developing baseline/counterfactual scenarios from which incremental costs and benefits can be measured and has conducted both retrospective and prospective technology assessments. Most of the projects he has led have involved modeling the adoption of new technologies and assessing the barriers to adoption. Dr. Gallaher has published books on research and development (R&D) in the service sector and the government's role in cyber security research. Dr. Gallaher has published articles in the *Journal of Technology Transfer*, *Journal of Energy Efficiency*, *International Journal of Technology Management*, *Environmental Sciences*, *International Journal of Production Economics*, and the *Journal of Law and Policy for the Information Society*. He has also made numerous presentations at professional conferences and to federal committees and industry associations, and has provided economic analysis support for Federal Advisory Committee Act (FACA) activities.

Albert N. Link is Professor of Economics at the University of North Carolina at Greensboro, USA. He is widely published in the fields of science and innovation policy, the economics of R&D, program evaluation, and productivity analysis. His research has appeared in academic journals such as the *American Economic Review*, *Journal of Political Economy*, *Review of Economics and Statistics*, *Economica*, *Research Policy*, and *Economics of Innovation and New Technology*. His most recent books include *Economics of Evaluation in Public Programs* with John Scott (Edward Elgar), *Public Goods, Public Gains: Calculating the Social Benefits of Public R&D* with John Scott (Oxford University Press), *Government as Entrepreneur* with Jamie Link (Oxford University Press), and *A History of Entrepreneurship* with Robert Hébert (Routledge). Professor Link is also editor-in-chief of the *Journal of Technology Transfer*. Professor Link has the privilege of being the US Representative to the United Nations

Economic Commission for Europe (UNECE), and in that capacity he is serving as the co-vice chairperson of the Team of Specialists on Innovation and Competitiveness Policies. He received a Ph.D. in economics from Tulane University.

Alan O'Connor is a senior economist in RTI's Environmental, Technology, and Energy Economics program. With nearly 15 years of experience in technology economics, public policy and regulatory impact analysis, and economic development, Mr. O'Connor specializes in leading interdisciplinary teams in economics studies at the interface between technology and economic, environmental, energy, and/or public health imperatives. His core research practice consists of cost-benefit analysis, technology program evaluations, industry surveys, policy analyses, and opportunity assessments for federal clients, including NIST, the DOE, the Centers for Disease Control and Prevention (CDC), and the EPA. Research objectives are often to determine program additionality, guide policymaking, or offer guidance on how public-sector investments and partnerships can simultaneously achieve economic and other goals. Mr. O'Connor also participates on RTI's divisional research teams for Public Health Services Environment, Health and Safety, Research Computing, and Innovation-Led Economic Development, lending his capabilities in economic impact modeling, primary data collection, partnership evaluation, and counterfactual scenario analysis. He has published in *The American Journal of Managed Care* and *The Economics of Innovation and New Technology*. Mr. O'Connor holds an MBA in international management with a concentration in strategic partnerships and finance.

Acknowledgments

We gratefully acknowledge the DOE's Office of Energy Efficiency and Renewable Energy (EERE) for funding the studies that provided the core research for this book. This study benefited from input from a number of individuals across the DOE, as well as other organizations. Jeff Dowd (DOE Study Project Manager) of EERE worked closely with Gretchen Jordan (Study Project Manager, Sandia National Laboratories) and Rosalie Ruegg (TIA Consulting, Inc.) to provide guidance and review of the analysis and results.

The solar technologies research benefited from the contributions of many individuals and organizations, especially the photovoltaic (PV) companies funded under multiple DOE initiatives that generously made their current and retired senior company executives and scientists available for interviews. In addition, many technologists and analysts were consulted, and the authors wish to thank several individuals in particular who reviewed draft material and, in some cases, participated in multiple rounds of interviews. Many thanks to Roland Hulstrom (National Renewable Energy Laboratory), Terry Jester (Hudson Clean Energy Partners), Sarah Kurtz (NREL), Jim McVeigh (Sentech Inc.), Rick Mitchell (NREL), Jim Rand (formerly of AstroPower), Ron Ross (formerly of Jet Propulsion Laboratory), Peter Sheldon (NREL), Arun Soni (Sentech Inc.), Scott Stephens (DOE), Harin Ullal (NREL), and Ken Zweibel (George Washington University).

Supporting the geothermal research, Doug Blankenship from Sandia National Laboratories provided insight into the history of polycrystalline diamond compact drill bits. Gregory Mines from Idaho National Laboratory provided information on the development of binary cycle power plant technology. Toshi Sugama provided insight into high-temperature geothermal well cements, and Karsten Prusess from Lawrence Berkeley National Laboratory provided information on the TOUGH series of reservoir models.

Supporting the vehicle technology research, James Eberhardt, Chief Scientist of the Vehicle Technologies Program (VTP) at EERE, helped define the scope and provided insight on the DOE's involvement in combustion research. Additional support was provided by Marvin Gunn, former Manager of the Combustion Research Program at DOE; William McLean, former Director of the Combustion Research Facility (CRF) at Sandia/Livermore; Gurpreet Singh, Team Leader of the Advanced Combustion Engine R&D subprogram within the VTP; and Dennis Sieberts, Manager of the Engine Combustion Research Program at the CRF.

Denise M. Mulholland from EPA was especially helpful in providing the code and documentation for EPA's COBRA model, which was used to monetize health benefits from reduced emissions.

We especially want to acknowledge and thank Ross Loomis, Fern Braun, Alex Rogozhin, Jeff Petrusa, Paramita Sinha, and Sara Casey from RTI for their research, analysis, and writing contributions to the DOE studies that formed the foundation of this book.

1. Introduction

> The [OPEC] Oil Embargo which began on October 19, 1973 sparked a fundamental reassessment of the nation's vulnerability to imported energy and also forced a reassessment of the role that energy R&D could play in helping secure the nation against hostile acts like the Oil Embargo. (Dooley, 2008, p. 9)

THE IMPORTANCE OF ENERGY TECHNOLOGY

The oil crisis of the early 1970s led the United States (and the world) to realize that cheap oil was not an 'inalienable right' and that the existing transportation system (from well head to wheel) was not sustainable. The decade also saw the Three Mile Island incident and witnessed the emergence of the environmental movement. By the end of the 1970s there was a consensus that new and more efficient technologies were needed to ensure economic growth, to reduce the United States's dependence on foreign energy sources, and to protect the environment.

Following the embargo, the US government aggressively began pursuing enhanced/alternative energy options. The DOE, activated on October 1, 1977, assumed the responsibilities of the Federal Energy Administration, the Energy Research and Development Administration, the Federal Power Commission and parts and programs of several other agencies, with the mission to '[e]nsure America's security and prosperity by addressing its energy, environmental, and nuclear challenges through transformative science and technology solutions.' (Energy.gov, 2011)

Over the past 30 years, a wide range of energy technology research programs have been pursued, under the frequently stated goal of the United States achieving energy independence within a specified number of years. To name a few, research programs have included technology development in areas of nuclear, solar, hydrogen, and biofuels. Over the decades, all of these initiatives have shown promise and to varying degrees been promoted as a potential (or at least partial) solution to energy issues with breakthrough on the horizon.

However, today we face much the same energy issues that were present in the 1970s. The United States's energy consumption distribution by energy source has changed only modestly (see Table 1.1). Whereas overall

Table 1.1 Comparison of US energy consumption by energy source

Energy source	1970[a] (% share of total)	1980[a] (% share of total)	2009[a] (% share of total)
Coal	13.0	16.3	20.9
Oil	31.2	36.2	37.3
Natural gas	23.0	21.4	24.7
Nuclear	0.3	2.9	8.8
Hydro	2.8	3.1	2.8
Geothermal	0.0	0.1	0.4
Wind	0.0	0.0	0.7
Solar	0.0	0.0	0.1
Biomass	1.5	2.6	4.1
Energy intensity per output [b] (GDP) (BTU/$)	15889	13379	7343
Energy intensity per capita [c] (Million BTU/Capita)	330.9	343.1	307.9

Sources: [a] EIA (2011), [b] BEA (2011), [c] Census (2011).

energy intensity has decreased (as measured as a share of gross domestic product [GDP]), little improvement has been made on energy consumption per capita. And the dependence on fossil fuel has persisted over time even in the face of increasing direct costs, and more recently, increasing concerns over the indirect costs associated with climate change and military interventions to ensure foreign supply.

It has become increasingly clear that in the near to midterm there is not likely to be a breakthrough innovation that on its own will solve the world's energy problems. For example, biofuels have many attractive attributes in that they are renewable and have the potential of low greenhouse gas (GHG) emissions. However, with the current technologies, production of 15 billion gallons of corn ethanol, which replaces only 12 per cent of imported oil, requires landmass of about the size of Iowa (Hertel et al., 2010).

In addition, problems in the intersection of energy production and food production are likely to increase over time as developing countries work to raise the standard of living of their growing populations.

Improvements are going to be needed across a broad range of energy technologies to meet the energy challenges of the 21st century. Research cannot just focus on a few high-profile, high-impact (but maybe low probability) energy initiatives. A true portfolio of energy technology research is needed and should also include low-tech solutions and incremental improvements to existing technologies. For example, improved insulation in commercial, residential, and institutional buildings has the potential to reduce US energy consumption by 15 per cent. As discussed throughout this book, combustion engines are likely to be the primary transportation power source for many years to come, and even small improvements in efficiency can lead to significant reductions in gasoline and diesel consumption.

There is a role for government investment in research throughout a technology's life cycle, and to its credit, the DOE has historically pursued a portfolio approach to energy technology development. However, much of the R&D public policy discussion has previously focused only on early-stage R&D, moving new technologies from laboratories to commercialization or developing radically new energy infrastructures. The case studies herein demonstrate that significant social returns can be achieved by removing barriers to enhancing and implementing existing technologies that are at different levels of maturity.

We present three case studies that focus on different technologies at different stages of maturity:

- primarily applied R&D (solar)
- market implementation (geothermal)
- efficiency improvements of a mature technology (combustion engine).

Each of these technological advancements has been developed in the private sector with technical and financial support from the DOE. As a starting point for the methodological development and case studies that follow, we begin with an overview of legislation related to developing energy technology.

LEGISLATIVE BACKGROUND

At the time of the Organization of the Petroleum Exporting Countries (OPEC) oil embargo, the US infrastructure related to energy was the Atomic Energy Commission (AEC). In response to the OPEC oil embargo, President Nixon launched Project Independence on November 7, 1973; the goal of the project was to achieve energy independence by 1980.

On December 4, 1973, President Nixon created the Federal Energy Office in the Executive Office of the White House to allocate then scarce petroleum supplies to refiners and consumers (Fehner and Holl, 1994).

More generally, in his State of the Union address on January 30, 1974, President Nixon stated: 'Let it be our national goal: At the end of this decade, in the year 1980, the United States will not be dependent on any other country for the energy we need to provide our jobs, to heat our homes, and to keep our transportation moving.' (American Presidency Project, 2011.)

On October 11, 1974, President Ford reestablished the Nixon emphasis on energy independence by signing the Energy Reorganization Act of 1974, Public Law 93-438. This Act built on the Federal Nonnuclear Energy Research and Development Act of 1974, which stated: 'The Congress declares the purpose of this Act to be to establish and vigorously conduct a comprehensive, national program of basic and applied research and development, including but not limited to demonstrations of practical applications, of all potentially beneficial energy sources and utilization technologies.' (Office of the Under Secretary for Defense, 2011.)

The Energy Reorganization Act established the Nuclear Regulatory Commission to carry out the responsibilities of the abolished AEC. The Act also created the Energy Research and Development Administration (ERDA) to, among other things, encourage and conduct: 'research and development in energy conservation, which shall be directed toward the goals of reducing total energy consumption to the maximum extent practicable, and toward maximum possible improvement in the efficiency of energy use [. . .] and research and development in clean and renewable energy sources.' (US Nuclear Regulatory Commission, 2002.)

Then, on August 4, 1977, President Carter signed the Department of Energy Organization Act of 1977, Public Law 95-91, transferring the mission of ERDA to the newly formed DOE. As stated in the Act, Congress finds that:

- The United States faces an increasing shortage of nonrenewable energy resources.
- This energy shortage and our increasing dependence on foreign energy supplies presents a serious threat to the national security of the United States and to the health, safety, and welfare of its citizens.
- A strong national energy program is needed to meet the present and future energy needs of the nation consistent with overall national economic, environmental, and social goals.
- Responsibility for energy policy, regulation, research, development, and demonstration is fragmented in many departments and agencies and thus does not allow for the comprehensive, centralized focus necessary for effective coordination of energy supply and conservation programs.

- Formulation and implementation of a national energy program requires the integration of major federal energy functions into a single department in the executive branch. (US Department of the Interior, 2011.)

By this Act, Congress declared that establishing such a department in the Executive Branch was in the public interest and would promote the general welfare by ensuring coordinated and effective administration of federal energy policy and programs. The DOE would, according to the Act:

> Carry out the planning, coordination, support, and management of a balanced and comprehensive energy research and development program, including—(A) assessing the requirements for energy research and development; (B) developing priorities necessary to meet those requirements; (C) undertaking programs for the optimal development of the various forms of energy production and conservation; and (D) disseminating information resulting from such programs.

The Office of Conservation and Solar Energy was created after the passage of the National Energy Conservation Policy Act of 1978, Public Law 95-619.

The office of EERE was formed within the DOE in 2001 when the Office of Conservation and Solar Energy was renamed and reorganized. The general goals of EERE are to advance a wide range of clean energy technologies, with the mission of strengthening the economy, protecting the environment, and reducing dependence on foreign oil. EERE's programs rely heavily on partnerships with the private sector, state and local government, DOE national laboratories, and universities.

EERE is organized around ten energy programs: Biomass Program; Building Technologies Program; Federal Energy Management Program; Geothermal Technologies Program (GTP, discussed in Chapter 6); Fuel Cell Technologies Program; Industrial Technologies Program; Solar Energy Technologies Program (discussed in Chapter 5); Vehicle Technologies Program (discussed in Chapter 7); Wind and Hydropower Technologies Program; and the Weatherization and Intergovernmental Program.

OVERVIEW OF THE BOOK

The remainder of the book is outlined as follows. In Chapter 2, the economic arguments for governmental investments in R&D and new technology are summarized. Therein is discussed the barriers to new technology that bring about market failure related to private investment in improved energy efficiency. The chapter emphasizes that these barriers exist throughout the technology life cycle. Associated with this involvement is a responsibility of accountability, as discussed in Chapter 3. Chapters 2 and 3 are brief, and although their subject matters are related, they are presented separately to emphasize the importance of the topics.

A counterfactual evaluation methodology is used to quantify the net social benefits of public investments in energy technology in the three case studies presented in Chapters 5, 6, and 7. Chapter 3 provides an overview and critique of this counterfactual evaluation methodology and related evaluation metrics.

Chapter 4 introduces the three retrospective case studies that follow. This chapter summarizes the technology background relevant to the case studies with an emphasis on applied R&D (investments in solar power), market technology (geothermal technology), and efficiency improvements in mature technology (combustion engines).

Chapter 5 presents an economic analysis of the net social benefits accruing from the investments of the DOE's Solar Energy Technologies Program in photovoltaic energy systems, specifically from photovoltaic module technologies that are encapsulated sets of solid-state solar cells that convert solar energy into electricity.

Chapter 6 presents the findings from an economic analysis of technology supported by the DOE's Geothermal Technologies Program. The study compares historical economic activity with the GTP's investments with what would have likely happened in the absence of these public investments.

Chapter 7 focuses on the Vehicle Technologies Program's investments in laser and optical diagnostics and combustion modeling for heavy-duty diesel engines. It describes how US diesel engine manufacturers have used the technology that came about from these public investments, and it offers quantitative measures of the resulting net social benefits.

Chapter 8 concludes the book with a brief statement of the policy implications to be drawn from the findings in the three case studies presented.

2. Economic rationale for public investment

GOVERNMENT'S ROLE IN INNOVATION

The theoretical basis for government's role in market activity is based on the economic concept of market failure. Market failure is typically attributed to market power, imperfect information, externalities, and public goods. The explicit application of market failure to justify government's role in innovation, and in R&D activity in particular, is a relatively recent phenomenon within public policy.

Many point in the United States to President George H.W. Bush's 1990 US Technology Policy (EOP, 1990) as that nation's first formal domestic technology policy statement. Albeit an important initial policy effort, it failed to articulate a foundation for the government's role in innovation and technology. Rather, it implicitly assumed that the government had a role and then set forth the general statement: 'The goal of US technology policy is to make the best use of technology in achieving the national goals of improved quality of life for all Americans, continued economic growth, and national security.' (EOP, 1990, p. 2.)

President William Clinton took a major step forward from the 1990 policy statement in his 1994 'Economic Report of the President' by articulating first principles about why government should be involved in the technological process: 'The goal of technology policy is not to substitute the government's judgment for that of private industry in deciding which potential "winners" to back. Rather, the point is to correct market failure.' (EOP, 1994, p. 191.)

President Clinton's 2000 'Economic Report of the President' elaborated on the concept of market failure as part of US technology policy: 'Rather than support technologies that have clear and immediate commercial potential (which would likely be developed by the private sector without government support), government should seek out new technologies that will create benefits with large spillovers to society at large.' (EOP, 2000, p. 99.)

MARKET FAILURE AND BARRIERS TO INNOVATION

Market failure (technological or innovation market failure, in particular) results from conditions that prevent organizations from fully realizing or appropriating the benefits created by their investments. To explain, consider a marketable technology to be produced through an R&D process where conditions prevent full appropriation of the benefits from technological advancement by the R&D-investing firm. Other firms in the market or in related markets will realize some of the profits from the innovation, and of course consumers will typically place a higher value on a product than the price paid for it. Then, because of such conditions, the R&D-investing firm will calculate that the marginal benefits it can receive from a unit investment in such R&D will be less than could be earned in the absence of the conditions, reducing the appropriated benefits of R&D below their potential, namely the full social benefits. Thus, the R&D-investing firm might underinvest in R&D, relative to what it would have chosen as its investment in the absence of the conditions. Alternatively stated, the R&D-investing firm might determine that its private rate of return is less than its private hurdle rate (i.e., the firm's minimum acceptable rate of return); therefore, it will not undertake socially valuable R&D.

A number of nonmutually exclusive factors can explain why a firm will perceive that its expected rate of return will fall below its hurdle rate. Eight are listed in Table 2.1 and discussed briefly below.

Table 2.1 Factors creating barriers to innovation that lead to technological market failure

1. High technical risk associated with the underlying R&D
2. High capital costs to undertake the underlying R&D
3. Long time to complete the R&D and commercialize the resulting technology
4. Underlying R&D spills over to multiple markets and is not appropriable
5. Market success of the technology depends on technologies in different industries
6. Property rights cannot be assigned to the underlying R&D
7. Resulting technology must be compatible and interoperable with other technologies
8. High risk of opportunistic behavior when sharing information about the technology

Sources: Link and Scott (1998, 2005, 2011).

1. High technical risk means the outcomes of the firm's R&D might not be technically sufficient to meet its needs. This might cause market failure, given that when the firm is successful, the private returns fall short of the social returns. An underinvestment in R&D will result.
2. High technical risk can relate to high commercial or market risk, when the requisite R&D is highly capital intensive. Such investments could require too much capital for a firm (any firm) to justify the outlay; thus, the firm will not make the investment, even though it would be better off if it had, and so would society.
3. Many R&D projects are characterized by a lengthy time interval until a commercial product reaches the market. The time expected to complete the R&D and the time until commercialization of the R&D results are long; thus, the realization of a cash flow is distant.
4. It is not uncommon for the scope of potential markets to be broader than the scope of the individual firm's market strategies, so the firm will not perceive economic benefits from all potential market applications of the technology.
5. The evolving nature of markets requires investment in combinations of technologies that, if they existed, would reside in different industries that are not integrated. Because such conditions often transcend the R&D strategy of individual firms, such investments are not likely to be pursued.
6. A situation can exist when the nature of the technology is such that it is difficult to assign intellectual property rights.
7. Industry structure can raise the cost of market entry for applications of the technology.
8. Situations can exist where the complexity of a technology makes agreement, with respect to product performance between buyers and sellers, costly.

With the government's involvement in the innovation process, which involves the use of public money, comes government's responsibility to be accountable for its fiscal actions.

3. Evaluation of public investments in new technology

ECONOMETRIC-BASED METHODOLOGIES

One set of evaluation methodologies that are used to quantify the economic impact of public investments in R&D (and we are assuming that R&D leads to new technology) is often based on econometric models, or more specifically one set is based on models that are estimated using econometric methods.[1] These models are not independent of the data that are available. Generally, the data, when available, pertain to the unit's (hereafter the unit is referred to as a firm for consistency in the discussion that follows) performance before and after the public sector R&D program.

To illustrate, define a performance variable for the ith firm as P_i. Consider two series of data. The first is a time series of data on the observed performance of k firms, $i = 1$ to $i = k$, before and after the effect from the public sector R&D program. After the public sector R&D program is operating, each of the k firms will be affected but not necessarily to the same degree. If performance data are available from time $t = 0$ to $t = n$, and if the public sector R&D program became effective at $t = t^*$, then the relevant comparison is between the performance of the k firms before the R&D program, P_i, for $t = 0$ to $t = t^*$, and their performance after the R&D program, P_i, for $t = t^*$ to $t = n$.

The second series of data could be on the performance of affected and nonaffected firms (i.e., matched pairs of firms) after the public-sector program was initiated at t^*. If performance data are only available from time $t = t^*$ to $t = n$ for k affected firms, P_i, $i = 1$ to $i = k$; and for m nonaffected firms, P_j, $j = 1$ to $j = m$; then, for each matched pair of firms, the relevant comparison is over time between P_i and its matched P_j.[2] The counterfactual situation, that is the situation proxied to be without the public sector R&D program, is the performance of the m nonaffected firms.

In the case of the first series of data that is quantifying pre- and post-R&D program performance, cross-sectional time series data could be used to estimate a model that takes the general form:

$$P_{i,t} = a_0 + a_1\, RD_{t^*} + \text{control variables} + \varepsilon, \tag{3.1}$$

where $P_{i,t}$ represents the relevant performance variable of the ith firm at time t; RD represents the public sector R&D program being evaluated that was initiated or became effective at time t^*—RD takes on a value of 0 for the time period before t^* and a value of 1 at t^* and afterwards;[3] and ε is a normally distributed random error term.

Estimated regression parameters from equation (3.1) allow one to interpret the economic impact of the public sector R&D program. For example, the estimated coefficient on RD in equation (3.1) quantifies the impact of the R&D program on the average performance of the sample of k firms. If the estimated value of a_1 is positive and statistically significant, then the public sector R&D program had a measurable positive impact on firm performance, all other factors held constant.

In the case of the matched pairs of firms in which the counterfactual situation is approximated by the performance of non-R&D program-affected firms, cross-sectional time series data could be used to estimate a model that takes the general form:[4]

$$P_t = b_0 + b_1\, E + \text{control variables} + \varepsilon. \tag{3.2}$$

The variable E divides the sample of firms into those affected by the R&D program and those matched pairs that are not affected—E takes on a value of 0 for the m nonaffected matched firms and a value of 1 for the k affected matched firms. If the estimated value of b_1 is positive and statistically significant, then the R&D program had a measurably positive impact on firm performance relative to the performance of 'similar' firms not affected by the program, other things held constant.

A number of important data issues are related to using the above econometric models, and one of those issues relates to how the performance variable, P, is measured. If P is measured in terms of the stated goals of the public sector R&D program, then the use of econometric models might be an appropriate tool for an economic assessment. If, however, P is measured in terms of performance of economic units beyond which the public sector R&D program was designed or focused (e.g., the overall performance of a sector or an industry although the R&D program was designed to affect firms, that is, spillovers have occurred), then the use of econometric models might be appropriate as an economic impact analysis tool.

Although evaluatory conclusions that are drawn from such analyses are useful, they are, in our opinion, incomplete in the sense that they do not consider the public sector's investment costs in R&D. We believe economics-based evaluation methodologies are more appropriate.

ECONOMICS-BASED METHODOLOGIES

Griliches (1958) pioneered the application of fundamental economic insight to the development of estimates of private and social rates of return to public investments in R&D.[5] Streams of investment outlays through time (the costs) generate streams of economic surplus through time (the benefits). Once identified and measured, these streams of costs and benefits are used to calculate rates of return, benefit-to-cost ratios, and other related metrics (e.g., net present value).

In the simplest Griliches model, public-sector innovations are conceptualized as reducing the cost of producing a good sold in a competitive market at a constant long-run unit cost. For any period, there is a demand curve for the good and, in the simplest model, a horizontal supply curve. Innovation lowers the unit cost of production, shifting downward the horizontal supply curve and thereby, at the new lower equilibrium price, resulting in greater consumer surplus (the difference between the price consumers would have been willing to pay and the actual price they paid, summed over all purchases).[6]

The Griliches model for characterizing the benefits from a public sector innovation has long been the traditional methodology in economics to follow for analyzing public sector R&D programs. The Griliches model for calculating economic social rates of return adds the public and the private investments through time to determine social investment costs, and then the stream of new economic surplus generated from those investments is the benefit. Thus, the evaluation question that can be answered from such an economics-based methodology is: What is the social rate of return to the innovation, and how does it compare with the private rate of return?

In practice, the stream of new economic surplus that is generated from the public sector investment is approximated in terms of the social benefits that exist with the public sector R&D and, counterfactually, without the public sector R&D. This so-called counterfactual approach underlies the evaluation methodology in the three case studies discussed in Chapters 5, 6, and 7. Integral to the analysis are the evaluation metrics used to assess the return to public sector R&D. These metrics are described in the next section.

EVALUATION METRICS

Internal Rate of Return

The internal rate of return (*IRR*) (a real rate of return in the context of constant-dollar cash flows) is the value of the discount rate, *i*, that equates the net present value (NPV) of the stream of net benefits associated with a research project to zero. The time series runs from the beginning of the research project, $t = 0$, through a terminal point, $t = n$.

Mathematically:

$$NPV = [(B_0 - C_0) / (1 + i)^0] + \ldots + [(B_n - C_n) / (1 + i)^n] = 0, \quad (3.3)$$

where, $(B_t - C_t)$ represents the net benefits associated with the project in year t, and n represents the number of time periods (years in the case studies of chapters 5, 6, and 7) being considered in the evaluation.

For unique solutions for i, from equation (3.3), the IRR can be compared with a value, r, which represents the opportunity cost of funds invested by the technology-based public institution. Thus, if the opportunity cost of funds is less than the internal rate of return, the project was worthwhile from an *ex post* social perspective.

Benefit-to-cost (B/C) Ratio

The ratio B/C is the ratio of the present value of all measured benefits to the present value of all measured costs. Both benefits and costs are referenced to the initial time period, $t = 0$, when the project began as

$$B/C = [\Sigma_{t=0 \text{ to } t=n} B_t / (1 + r)^t] / [\Sigma_{t=0 \text{ to } t=n} C_t / (1 + r)^t]. \quad (3.4)$$

A B/C ratio of 1 is said to indicate a project that breaks even. Any project with $B/C > 1$ is a relatively successful project as defined in terms of benefits exceeding costs.

Fundamental to implementing the ratio of benefits to costs is a value for the discount rate, r. The rate emphasized in the case studies herein follows the guidelines set forth by the Office of Management and Budget (OMB) (1992) in Circular A-94: 'Constant-dollar benefit-cost analyses of proposed investments and regulations should report net present value and other outcomes determined using a real discount rate of 7 per cent.'[7]

Net Present Value

OMB circular A-94 states:

> The standard criterion for deciding whether a government program can be justified on economic principles is net present value—the discounted monetized value of expected net benefits (i.e., benefits minus costs). Net present value is computed by assigning monetary values to benefits and costs, discounting future benefits and costs using an appropriate discount rate, and subtracting

the sum total of discounted costs from the sum total of discounted benefits. (OMB, 1992, p. 3)

The information developed to determine the B/C ratio can be used to determine NPV as

$$NPV_{\text{initial year}} = B - C,$$

whereas in the calculation of B/C, B refers to the present value of all measured benefits and C refers to the present value of all measured costs, and where present value refers to the initial year or time period in which the project began, $t = 0$ in terms of the B/C formula in equation (3.3). Note that NPV allows, in principle, one means of ranking several projects *ex post*, providing investment sizes are similar.

NOTES

1. It is important to emphasize the difference between the terms 'methodology' and 'method'. The terms are often used interchangeably, although it is incorrect to do so. A methodology is the theoretical foundation or practices within a discipline that determine or guide how to engage in an inquiry; a method is a tool or technique used to implement the inquiry.
2. Many issues are related to how one defines a matched firm (e.g., size, level of own R&D, industry), but that discussion is beyond the scope of this book.
3. The specification in equation (3.1) is simplistic in the sense that it assumes the R&D program's impact occurred at time t^* and that the impact remained constant through subsequent time periods. More sophisticated variations of equation (3.1) are possible.
4. A number of econometric issues are associated with the models below that are not discussed herein, such as participation in the R&D program denoted by *RD* in equation (3.2), but that could be relevant to an actual study.
5. The Mansfield *et al.* (1977) seminal article applied the Griliches methodology to private-sector innovations.
6. Additionally, for market settings more complicated than the simplest model, the Griliches model accounts for producer surplus, measured as the difference between the price the producers receive per unit and the actual marginal cost, summed over the output sold, minus any fixed costs. Social benefits are then the streams of new consumer and producer surpluses—economic value above and beyond the opportunity costs of the resources used to create value, while private benefits for a firm that invests in innovation are the portions of the streams of producer surplus appropriated by the investor. Not all of the appropriated producer surplus is necessarily new because the surplus gained by one producer might be cannibalized from the preinnovation surplus of another producer. Social and private costs will, in general, also be divergent.
7. The three case studies in Chapters 5, 6, and 7 evaluate investment decisions—the allocation of capital across alternative investment options, and so the OMB-mandated 7 per cent real discount rate has been used. Commenting on the 7 per cent real discount rate, OMB (2003, p. 33) observed: 'The 7 per cent [real] rate is an estimate of the average before-tax rate of return to private capital in the US economy. It is a broad measure that reflects the returns to real estate and small business capital as well as corporate capital. It approximates the opportunity cost of capital, and it is the appropriate discount rate

whenever the main effect of a regulation is to displace or alter the use of capital in the private sector. OMB revised *Circular A-94* in 1992 after extensive internal review and public comment.' Further, OMB (2003, p. 33) observed: 'The pre-tax rates of return better measure society's gains from investment. Since the rates of return on capital are higher in some sectors of the economy than others, the government needs to be sensitive to possible impacts of regulatory policy on capital allocation.' However, OMB (2003, p. 33) states: 'The effects of regulation do not always fall exclusively or primarily on the allocation of capital. When regulation primarily and directly affects private consumption (e.g., through higher consumer prices for goods and services), a lower discount rate is appropriate.' Hence, if one were evaluating a policy where, instead of alternative uses of investment capital in public R&D investment decisions, the issue evaluated were a regulatory policy (e.g., for health care) that would directly and primarily affect the stream of real income to consumers (e.g., alternative health plans with streams of different magnitudes and different timings), then OMB has directed (2003, pp. 33–34) that 'for regulatory analysis' (p. 34), rather than an evaluation of an investment, the real discount rate of 3 per cent should be used and then compared with the results using the 7 per cent real discount rate. OMB explains that for consumers's decisions, 3 per cent better approximates their real rate of time preference.

OMB (2003, p. 33) explicitly stated that the 7 per cent real required rate of return that is based on the average rate of return to private capital investment is 'a default position,' yet, the market failure story recognizes that for investments (not just 'regulatory policy') the social rate of return and the private rate of return can (and are expected to) diverge, with the social required rate of return being less than the private hurdle rate. As it turns out, in practice, a 7 per cent social hurdle rate for public investments is not inconsistent with that logic, because the 7 per cent is based on the average, but for the R&D investment projects we evaluated in the case studies the firms report higher private hurdle rates. OMB appears to be taking the least controversial approach by using for the social hurdle rate for investments an average return for private capital investments and by advising consideration of the variance in private returns in different activities. Clearly, as we have noted, there is no reason society should be constrained to its assessments of value by prices determined in markets where there are market failures and the prices give the wrong signals. Hence, the private rate of return on investment should not be expected to equal the social opportunity cost of investment funds; the private rates of return may be based on prices that do not reflect social value. We know that with positive externalities such as nonappropriated spillovers that benefit those who did not invest, social rates of return can be high when private rates of return are low. Moreover, the private rate of return can be high even when the social rate of return is low or even negative. For example, in the context of R&D investment, the results of a privately profitable R&D investment may simply cannibalize previously existing economic surplus, causing the investment to have a negative social rate of return. OMB's approach is a solution in the absence of a practical way to determine what the theoretical social hurdle rate should be in any given situation.

4. Technical discussions of the case studies

INTRODUCTION

The technical backgrounds for the three case studies that follow are discussed in this chapter. It is important to compare and contrast these backgrounds to appreciate the scope of research (e.g., applied R&D, market technology, and improvements in mature technology) that EERE supports and that is evaluated in the following chapters.

EERE is currently organized around the ten technology programs listed in Table 4.1. Table 4.1 also provides a general assessment of the life-cycle stage of the technology programs and their status related to commercialization and existing markets. It should be noted that most of the technology programs individually pursue a portfolio of research that spans from applied research and development to deployment of mature technologies.

As mentioned previously, EERE's programs rely heavily on partnerships with the private sector, state and local government, DOE national laboratories, and universities. This wide range of public sector and private sector partnerships reflects the variation of where the research is relative to the technology life cycle. For example, for early biomass technologies, EERE partners heavily with universities and national laboratories. For more mature technologies, as is the case with wind and water power technologies, EERE partners with private-sector companies to make incremental improvements. For mature and commercially available technologies, EERE works with federal, state, and local government entities to adoption.

The three case studies included in the following chapters document the range of research activities conducted by EERE at different stages of the technology life cycle:

- The Solar Energy Technologies Program (SETP) focused on applied R&D targeted at developing the technological breakthroughs needed to make large-scale solar power cost-competitive and penetrate beyond the few niche areas where it historically had been viable. Among other activities, SETP funds technical infrastructure, technical R&D, and production scale-up.

Table 4.1 Technology life cycle stage of EERE's programs

Current EERE program	Technology life cycle stage
Biomass Program	Applied research, development, and market implementation
Building Technologies Program	Applied research, development, and market implementation
Federal Energy Management Program	Market implementation
Geothermal Technologies Program	Applied research, development, and market implementation
Fuel Cell Technologies Program	Applied research, development, and market implementation
Industrial Technology Program	Applied research, development, and market implementation
Solar Energy Technologies Program	Applied research, development, and market implementation
Vehicle Technologies Program	Applied research, development, and market implementation
Wind and Water Power Technologies Program	Market implementation
Weatherization and Intergovernmental Program	Applied research, development, and market implementation

- The Geothermal Technologies Program (GTP) worked to solve implementation problems and demonstrated that existing technologies were reliable and economical, growing the confidence the power industry needed for large-scale implementation.
- The Vehicle Technologies Program (VTP) worked with automakers to pursue efficiency improvements to existing combustion technology that the auto industry would not have been able to pursue without EERE's modeling expertise.

The three case studies provide a snapshot of how EERE engages in technology development through the life cycle of energy technologies. The key point is that in addition to the importance of aggressively pursuing R&D that will enable the next generation energy infrastructure is the necessity to pursue near-term solutions and enhancements to existing and mature energy technologies. Long-term energy needs must be balanced with short-term energy needs. For sustainable growth, R&D should not focus solely on the next-generation technology, nor should it focus solely on the development and deployment of this generation's technologies.

The three case studies document the economic impact resulting from selected EERE efforts: applied R&D (SETP), market technology (GTP), and improvements in mature technology (VTP). In all three cases, significant energy savings and social return were realized on taxpayers' investment.

Three technical overviews follow. We have intentionally segmented the technical discussion of the underlying technologies from the evaluation analyses, which are in separate chapters, in anticipation of the differing interests of readers with various backgrounds.

SOLAR TECHNOLOGIES

The solar technology case study is a retrospective analysis of net benefits accruing from DOE's investment in photovoltaic (PV) technology development. The study employed a technology cluster approach. That is, benefits measured for a subset of technologies in a meaningful cluster, or portfolio, of technologies were compared with the total investment in the cluster to provide a lower-bound measure of return for the entire cluster.

The technologies selected for analysis were PV module technologies. PV modules are encapsulated sets of solid-state cells that convert solar energy into electricity. They are perhaps most recognizable as the flat-plate solar panels mounted on rooftops, affixed to signal posts, or assembled in large arrays that compose solar farms. PV modules are usually characterized by the material technologies that compose the cells. These may be crystalline silicon (c-Si) or 'thin films' of semiconductor material such as cadmium telluride (CdTe) or copper indium diselenide (CIS).

The technology cluster was Photovoltaic Energy Systems, which is one of the four thrusts within the DOE Solar Energy Technology Program (SETP).[1]

PV technologies have benefited from long-term DOE investment in core cell and module technology R&D, manufacturing process development, and the technology infrastructure enabling that R&D. Between 1975 and 2008, the period of analysis for this case study, researchers from industry, academia, and DOE's national laboratories received financial and technical support to hasten the development and market introduction of higher-quality, longer-lived and lower-cost PV modules.

There has been a national solar energy imperative since the beginning of the OPEC oil embargo in 1973, which led to an immediate concern about energy security in the United States. Coincidentally, the National Science Foundation (NSF) and the National Aeronautics and Space Administration (NASA) had been planning a conference to lay out funding and develop a plan for terrestrial PV development. At the time, the domestic PV industry was in its infancy, and technical expertise was concentrated at NASA's Jet Propulsion Laboratory (JPL), which developed photovoltaics for space applications. Referred to as the Cherry Hill Conference, this conference was held just one week after the oil embargo began, giving it great national significance.

The Cherry Hill Conference established technology goals for terrestrial photovoltaics and marked the beginning of the National Photovoltaics Program. The following year, after the creation of the ERDA (the precursor to DOE), the Solar Energy Research, Development, and Demonstration Act called for research and commercialization programs and established the Solar Energy Research Institute (now the National Renewable Energy Laboratory [NREL]), which began operation in 1977. In the years that followed, DOE deployed long-term, sustained R&D initiatives that were responses to technical barriers or technology opportunities for terrestrial photovoltaics.

Three of these initiatives are of particular focus in this analysis, each of which was a broad technology response to the technical barriers and technology needs present at the time the initiative was launched, building on the technology base developed by its predecessor:

- The Flat-Plate Solar Array (FSA) Project (1975–1985), which was funded by ERDA and DOE but managed by JPL in order to transfer JPL's rich space-based PV expertise to the nascent terrestrial PV industry and the Solar Energy Research Institute.
- The Photovoltaic Manufacturing Technology (PVMaT) Project (1991–2008), which was later renamed the Photovoltaic Manufacturing Research and Development Program and under which advanced manufacturing technologies for cell production and module assembly were developed to hasten cost reductions and improve efficiency, quality, and reliability.
- The Thin-Film PV Partnerships (TFP) Program (1994–2008), which was preceded by the Amorphous Silicon and Polycrystalline Thin-Film programs dating back to the 1970s and aimed to develop thin-film PV technologies.

FSA aggressively targeted core reliability, quality, and efficiency barriers to move photovoltaics from niche off-grid applications to the mainstream.

Industry experts interviewed for this study universally regarded the FSA period as foundational to the modern terrestrial PV industry. In 1975, the US PV industry produced 0.4 megawatts (MW) at a production cost per watt of $83.86 (2008$). Each module produced had no warranty and was expected to have a useful life of two to three years. When the FSA officially ended in 1985, 7.8 MW (+2,000 per cent) were produced at a production cost per watt of $9.40 (−82 per cent), and ten-year warranties were offered (Table 4.2).

End-year FSA milestones were largely the results of technology developed by 1983 and 1984, and industry progress slowed during most of the 1980s after federal funding for technology development was reduced. PVMaT was launched in 1991 to reinvigorate progress by developing manufacturing technologies. Despite progress under the FSA, many production processes remained manual. Furthermore, FSA had identified thin films as a viable alternative to c-Si, but little commercialization progress had been made. TFP would focus on bringing thin-film technologies to commercialization, and PVMaT would develop the manufacturing technology to increase operational efficiencies through process development and automation. In 1991, the US PV industry produced 17.5 MW at a production cost per watt of $6.93 (2008$). In 2008, 1,022.6 MW (+>5700 per cent) was produced at a production cost per watt of $1.92 (−72 per cent). Over 60 per cent of 2008's production volume was in thin-film PV modules.

Flat-Plate Solar Array Project (FSA)

Commercially available PV modules in the early to mid 1970s had low efficiency ratings in the range of 4.8 to 6.5 per cent, were priced between $80 and $150 per watt (2008$), had no warranty, and were largely unimpressive (Christensen, 1985; Green, 2005). The Cherry Hill Conference called for developing the entire technology base that would bring PV from a curiosity or niche market application into the mainstream and ultimately into grid-connected systems.

The Massachusetts Institute of Technology's (MIT's) Energy Laboratory supplied the DOE with an assessment of the nascent terrestrial PV industry and provided the public policy analysis framework for guiding public investment in PV (Linden et al., 1977). Linden et al. explored the interplay between technology development, production, and public policy to overcome market failures and technical obstacles. The report identified the primary market failures that were inhibiting the development of terrestrial photovoltaics:

Table 4.2 US PV industry progress, 1976–2008

Year	Module production (MW)			Production Cost ($/W)	Reliability (years)	Notable technology outcomes
	c-Si	Thin films	Total			
1974	0.19	0.00	0.19	114.44	2	**Flat-Plate Solar Array Project** Block Purchases I-V; EVA for encapsulants; UCC silicon refining process; silicon ingot growth; silicon ribbon growth, automated module assembly; design and test methods for durability; performance, and safety; laboratory cells reaching 22% efficiency; and 10-year module warranties
1975	0.37	0.00	0.37	83.86	2	
1976	0.80	0.00	0.80	53.28	2	
1977	1.22	0.00	1.22	37.60	2	
1978	1.65	0.00	1.65	25.64	2	
1979	2.07	0.00	2.07	23.93	2	
1980	2.50	0.00	2.50	22.22	2	
1981	4.46	0.00	4.46	19.65	2	
1982	5.05	0.00	5.05	17.09	5	
1983	5.63	0.00	5.63	14.53	5	
1984	6.22	0.05	6.27	11.96	5	
1985	7.30	0.50	7.80	9.40	10	
1986	6.40	0.85	7.25	8.99	10	
1987	7.45	1.40	8.85	8.58	10	
1988	9.70	1.85	11.55	8.16	10	
1989	12.95	1.45	14.40	7.75	10	
1990	13.78	1.37	15.15	7.34	20	
1991	16.48	1.00	17.48	6.93	20	**Thin-Film PV Partnerships** National teams; basic research in a-Si, CdTe, and CIS; a-Si modules (ECD/Uni-Solar); CdTe modules (First Solar [Solar Cells Inc.]); and CIS/CIGS modules (Global Solar)
1992	16.95	1.65	18.60	6.00	20	
1993	20.91	1.53	22.44	5.69	20	
1994	24.31	1.95	26.26	4.84	20	
1995	33.30	1.66	34.96	4.53	20	
1996	37.35	2.46	39.81	3.93	20	
1997	48.00	3.10	51.10	3.77	25	

Table 4.2 (continued)

Year	Module production (MW)			Production Cost ($/W)	Reliability (years)	Notable technology outcomes
	c-Si	Thin films	Total			
1998	48.10	5.80	53.90	3.71	25	
1999	53.80	7.00	60.80	3.45	25	**PV Manufacturing Technology Project** Wire saw technology adoption for silicon ingot wafering; automated cell and module assembly processes; in-line diagnostics and monitoring; high-efficiency c-Si cells; cost reductions from $6.93 per watt in 1991 to $1.92 per watt in 2008; 25-year module warranties; funded AstroPower (GE), BP Solar (Solarex), Evergreen, First Solar, Global Solar, SCHOTT Solar, SolarWorld USA (Arco/Siemens/ Shell), SunPower, Uni-Solar
2000	66.00	9.00	75.00	2.96	25	
2001	86.70	13.80	100.50	3.00	25	
2002	109.40	18.20	127.60	2.85	25	
2003	86.82	15.80	102.62	2.91	25	
2004	115.20	23.50	138.70	2.80	25	
2005	133.60	44.50	178.10	2.96	25	
2006	175.30	92.50	267.80	2.67	25	
2007	189.20	263.00	452.20	2.11	25	
2008	379.90	642.70	1 022.60	1.92	25	

Note: EVA = ethylene vinyl acetate.

Sources: Christensen (1985); (Maycock, 1986–2004; *PV News*, 2005–2009; EIA and IEA (EIA, 2008; IEA, 2009); Friedman et al., 2005; Green (2005).

- Incorrect energy prices that do not account for deleterious environmental or human health impacts associated with fossil fuel consumption and combustion.
- Production uncertainties concerning prices, availability, quality,

reliability, production volumes, and the ready supply of renewable and fossil fuel technology alternatives.
- Technological uncertainties, particularly with respect to development costs, time, and R&D performance.
- Interdependencies of production and technology development, which are the confluence of uncertainties, indivisibilities, and externalities that impede market function through asymmetries in information and poor convergence of expectations.
- Indivisibilities and inability to appropriate returns from technology development, so that, despite photovoltaics being in the national interest, the costs of developing and maturing the technology may preclude private-sector innovation if returns from innovation cannot be appropriated as profits within a suitable time horizon.
- Imperfections in financial markets attributable to the chasm between internal sources of funding and the risk–reward profile that influences private equity financing.
- Noncompetitive market structures that may inhibit new, competing sector development.

In response, the funding for photovoltaics in the early years addressed both the supply side and the demand side of technology development. Industry, university, and government researchers established two major goals that drove FSA's mission to lower costs, increase efficiency, and increase reliability. The first was to demonstrate technologies that, if scaled to commercial production levels, could achieve a module production price of $1.62/Wp (2008$) with a 10 per cent efficiency and 20-year lifetime. The second was to mass-produce this technology.

The DOE funded applied research within the industry to improve c-Si module design and production technology and acted as the primary purchaser of these products. These purchased PV products were then tested by FSA researchers, and companies used the test results to improve their products. Funding was from ERDA, but JPL was selected to manage the project, given its extensive expertise in developing photovoltaics for space applications. Previous spaceflight projects provided JPL staff with invaluable experience in reliability testing and technical skills that were not available elsewhere. The US government's initial interest in developing PV technology was for space applications, with solar cells used to power a backup radio transmitter in the Vanguard I satellite in 1958 (Margolis, 2002). PV technology, although expensive, did not represent a large portion of the costs associated with NASA's programs, and applied R&D at JPL focused on improving the technology for space applications without great regard to its cost. In 1970, the average cost of space

PV modules was about $150 per watt (1970$) (Margolis, 2002). ERDA planned the launch of the Solar Energy Research Institute and the development of programs at Sandia National Laboratories (SNL). Contracting with JPL offered an opportunity to transfer expertise between federal programs and the nascent PV industry.

FSA was originally organized in five sections: silicon material refinement, c-Si sheet formation, automated module assembly, encapsulation, and large-scale production. In 1982, a high-efficiency cell task was added. In addition to these technical tasks, FSA included a project analysis and integration area to integrate the other project areas, provide economic analyses, and assess technical progress. Periodic economic analyses were used to judge the potential of current technical options and cancel unpromising pathways.

Despite frequent redirections and funding cuts due to shifting national priorities, FSA had achieved many of its objectives when it ended (Christensen, 1985). Module prices were reduced by a factor of 15, and efficiencies for modules in commercial production increased from about 5 to 10 per cent. Reliability improvements sparked by testing at FSA allowed companies to offer at least ten-year warranties on modules, whereas before FSA, warranties were nonexistent in the PV industry.

Researchers had studied existing terrestrial PV systems and found that many of these systems failed within a year of installation and that no warranties were offered. Causes of module failure were rapidly understood and addressed through R&D collaboration between industry and government. Reliability technology was transferred efficiently to industry, and by the early 1980s, c-Si module manufacturers had converged on a module design that is essentially the same as it is today. One industry veteran noted that the PV industry stated that PV module 'failure rates [before FSA] were horrendous' and 'this early work was the best and has stood the test of time.' Core industry standards were established. Underwriters' Laboratories (UL) standards and International Electrotechnical Commission standards are traceable to FSA.

The actual, nominal-dollar investment in FSA between 1975 and 1985 was $228 million (Christensen, 1985). Annual expenditure data were adjusted to 2008 dollars, and in inflation-adjusted terms, the total investment was $535 million (Table 4.3).

Silicon Material Refinement

Abundant polysilicon feedstock is necessary for large-scale c-Si PV production, and the cost of polysilicon is a significant contributor to the total cost of c-Si PV modules. To realize their cost goals, FSA funded R&D for

Table 4.3 DOE expenditures for FSA

Fiscal Year	Nominal ($ thousand)	Deflator	Real 2008$ ($ thousand)
1975	600	0.31	1939
1976	11 700	0.33	35 765
1977	30 900	0.35	88 796
1978	31 800	0.37	85 390
1979	32 900	0.40	81 559
1980	30 500	0.44	69 291
1981	28 600	0.48	59 409
1982	16 700	0.51	32 694
1983	13 600	0.53	25 613
1984	15 000	0.55	27 227
1985	15 500	0.57	27 307
Total	227 800		534 990

Sources: Christensen, 1985; GDP Implicit Price Deflator (2005 = 100) from DoC (2009).

many technologies that had the potential to lower the cost of polysilicon feedstock relative to the existing manufacturing process involving silicon deposited in a Siemens-type reactor from trichlorosilane gas. FSA funded 11 different contractors, each with a unique vision for polysilicon production processes. The most successful was the silane-to-silicon process at the Union Carbide Corporation (UCC) using fluidized-bed reactors. The UCC process uses silane gas as opposed to trichlorosilane as a feedstock to deposit polycrystalline silicon using the Siemens process. Advantages of the UCC process include 'a lower deposition-reaction temperature, a higher conversion efficiency, and lower environmental and corrosion problems' (Lutwack, 1986). UCC demonstrated the ability to produce purified polysilicon from metallurgical-grade silicon at lower costs (Christensen, 1985).

Silicon Sheet Formation: Wafers and Ribbons

FSA explored three categories of sheet formation: ingot growth with subsequent wafering, ribbon growth, and silicon coating on a substrate. Although none of the silicon coating methods met cost, yield, or performance goals for the project, researchers were successful in reducing cost and increasing yield in the Czochralsi (Cz) ingot growth process. However, ingots must be sliced into wafers for use in PV cells, and the wafering process can be time-consuming and expensive, wasting large amounts of valuable polysilicon feedstock material. To address this problem, FSA evaluated several different wafering technologies, none of which met speed and yield goals. (This technical challenge would later be overcome during PVMaT when researchers successfully adopted wire-saw technologies.) Five ribbon growth methods were examined. High-throughput growth and multiple ribbon growth were achieved with ribbon growth using the edge-defined film-fed growth method (EFG). Mobil Solar demonstrated EFG's performance during FSA.

High-efficiency Solar Cells

The 1983 DOE Five-Year Plan set a goal of 15 per cent efficiency for low-cost modules, which would require production cells with over 17 per cent efficiency. High-efficiency research at FSA focused on reducing bulk losses in the silicon, reducing surface losses, improving design and production, and improving modeling and measurements to reach this goal. Conversion efficiency increased greatly during the years of the task, with laboratory cells reaching 22 per cent efficiency (Christensen, 1985).

Encapsulants

FSA explored encapsulant materials and processes to identify an encapsulant that could provide a 20-year module life at a low cost. The most significant accomplishment of this task was the development of improved ethylene vinyl acetate (EVA) suitable for mass module production. Prior to using EVA as a laminating material, modules used a polyvinyl butyral encapsulant or silicon rubber, both of which faced problems with exposure to the elements. EVA was commercialized through FSA and remains the standard encapsulant in modules 25 years later.

Process Development and Automated Module Assembly

More than 140 processes were developed and transferred between industry and government partners, including those for cell surface treatment, junction formation, metallization, and module fabrication). Modules on the market in 1975 suffered from labor-intensive processes, high material costs and low cell-packing factors (Gallagher et al., 1986). The process development thrust under FSA was formed to decrease the cost of module production through automation and development of manufacturing technologies. More than 75 contracts were issued in two groups: low-cost processes and high-efficiency cell processes. Research in this area led to the successful demonstration of robotic module assembly and resulted in many new processes and equipment.

Large-scale Production (Block Purchase Program)

During FSA, JPL, via its large-scale production thrust, was responsible for procuring and testing modules from large production runs and tested more than 150 different module designs (Christensen, 1985). Through its Block Purchase Program, JPL purchased and tested a series of five block purchases of modules, offering feedback to manufacturers. Manufacturers would attempt to fix the problems identified by JPL, perform R&D to overcome shortcomings, and submit modules for the next round of testing. Testing began with very primitive modules, which performed poorly and degraded quickly with exposure to the elements.

Modules improved so drastically from Block I (1976) to Block V (1984) that the modules evaluated in Block V were not significantly different from those used today. Module prices fell from $152/W in 1974 to $12.50/W in 1985 (2008$). Block I modules had an average lifetime of under three years with no warranty. Block V modules offered ten-year warranties, and the expected useful life of a module produced in 1985 was around 30 years.

As part of the qualification and testing process for block purchases, this work also established the technology infrastructure for efficiency measurement, materials characterization, and reliability testing, including the following:

- design and test methods for performance, environmental durability, and safety
- materials characterization and optimization methods
- module fabrication methods and system designs for durability, safety, and performance
- robust reliability physics, test methods, and equipment
- reference materials and US and international PV standards.

Photovoltaic Manufacturing Technology (PVMaT) Project

Whereas FSA aimed to rapidly develop technologies throughout the PV module value chain, PVMaT targeted manufacturing technologies that would enable PV companies to accelerate decreases in production costs and increases in capacity.[2] Despite all the gains from technologies developed under FSA, much assembly was still performed by hand, and technical challenges involving crystal growth, wafer slicing, deposition, encapsulation, and other issues made it difficult for companies to reduce costs or increase capacity.

PVMaT was also created in part as a response to the falling US share of the global PV market, which had dropped significantly in the years since the last block purchase under FSA in 1984 (Mitchell et al., 1998). The United States, once the world's only major producer of PV systems, saw a significant growth in competition from Japan during the 1980s. Although US government funding for PV R&D had declined during the 1980s, Japanese government funding was much higher than it was in the 1970s. PVMaT was envisioned as a way to ensure that the United States would remain a major competitor in the global PV market. A strong domestic PV industry could lead to job creation and correct trade imbalances while providing a source of renewable energy and increasing energy independence (Mitchell et al., 1998).

PVMaT's goals were to:

- improve module manufacturing processes and equipment[3]
- reduce the cost of manufacturing PV modules, balance-of-systems components, and integrated systems
- improve module performance and reliability
- increase US PV manufacturing capacity (Mitchell et al., 1998).

The DOE generated financial leverage for emergent PV companies through cost-sharing plans to accelerate manufacturing technologies and products in ways that otherwise might not have occurred or would have taken longer to materialize. PVMaT was technology neutral: All PV companies with viable strategies for improving their production technologies were invited to submit proposals for funding. Successful proposers would receive DOE cost sharing up to 50 per cent of the total project cost and preferential access to NREL and SNL technology experts.[4]

PVMaT was conducted in 11 phases between 1991 and 2008. Each phase was an R&D response to technical challenges facing the industry at the time the phase was conceived. Companies helped NREL identify the major technical issues the industry faced, and NREL developed a roadmap for the initiative. Projects were awarded by a panel of PV experts.[5] Although the largest portion of PVMaT contract funding went toward improving c-Si technologies, PVMaT also supported thin-film companies in the scale-up of their manufacturing processes. All eight major US producers of PV received PVMaT funding. Of those eight, seven were in the top ten recipients of PVMaT funds from NREL.

The following is a summary of notable accomplishments by phase (Table 4.4), many of which occurred early in PVMaT between 1992 and 1996 (Margolis, 2002):

Table 4.4 PVMaT phases

Phase	First year	Research focus	DOE cost share ($ thousands)	Company cost share ($ thousands)	Total ($ thousands)
1	1991	Problem identification	1 053	–	1 053
2A	1992	Problem solving: process-specific manufacturing	30 738	21 316	52 055
2B	1993	Problem solving: process-specific manufacturing	13 384	14 557	27 941
3A	1993	Problem solving: teamed research on generic problems	2 220	752	2 972
4A1	1994	Product-driven PV systems and component technology	5 343	1 812	7 155
4A2	1994	Product-driven PV module manufacturing	14 349	10 167	24 516
5A1	1998	PV system and component technology	4 261	4 700	8 961
5A2	1998	PV module manufacturing technology	26 451	20 689	47 140
IDIP-1	2001	In-line diagnostics and intelligent processing: PV system and component technology	3 593	3 807	7 400
IDIP-2	2001	In-line diagnostics and intelligent processing: PV module manufacturing technology	23 369	30 443	53 812
YDR-1	2003	Large-scale module and component yield, durability and reliability	2 860	6 358	9 219
YDR-2	2003	Large-scale module and component yield, durability and reliability	23 397	23 773	47 170
Total			151 018	138 375	289 393

Note: Dollars are presented in nominal terms. IDIP = in-line diagnostics and intelligent processing. YDR = yield, durability, and reliability. Dollar values presented exclude DOE program administration expenditures.

Source: NREL (2009b).

- Phase 1 was an exploratory phase under which all US PV companies were invited to receive planning grants of up to $50,000 to study and recommend ways in which their processes could be improved to meet PVMaT goals.
- Phases 2A and 2B focused on process-specific problem solving and were directly related to low-hanging fruit identified during Phase 1. One of the most significant accomplishments of Phase 2 was Solarex's successful adoption of the wire saw, a technology that reduced silicon waste, increased wafer size, and that would later be adopted across the semiconductor industry. Solarex (now BP Solar) had explored the use of wire saws before PVMaT but had been unable to obtain funding to successfully implement them in their production process (Margolis, 2002).
- Phase 3A emphasized teamed research for generic, industry-wide problems. Teams consisted of combinations of university and industry partners. Spire Corporation, with Solec International and University of Massachusetts automation specialists, developed improved automated cell assembly processes that had lower costs. Springborn, with other companies and universities acting as subcontractors, developed new EVA encapsulants that resolved discoloration and degradation issues.
- Phases 4A1 and 4A2 focused on product-driven, full-system issues. Phase 4A was split into two parts to address system components (Phase 4A1) and module manufacturing (Phase 4A2) separately. Ascension Technology and ASE Americas developed an alternating-current module. AstroPower also created the world's largest production silicon solar cell, a record efficiency 1 cm^2 cell and a high-speed silicon-film production process (NREL, 2009d).
- Phases 5A1 and 5A2 continued the R&D trajectory set by Phase 4. Crystal Systems Inc. successfully designed a process to convert metallurgical-grade silicon to solar-grade silicon, reducing the cost of solar-grade silicon to less than $10 per kilogram—a price much lower than the contract goal. In Phase 5A2, BP Solar created a fully automated high-throughput cell-processing system (NREL, 2009d).
- Phases IDIP-1 and IDIP2 were designed to increase yield-of-module and BOS components through improved in-line diagnostics and monitoring. Sinton Consulting developed an in-line monitoring tool that allowed low-quality materials to be removed before becoming a cell. Evergreen Solar developed its string-ribbon silicon growth process and successfully moved a dual-ribbon growth system from R&D to production while drastically increasing throughput (NREL, 2009d).

- Phase YDR was intended to increase yield and reliability through better PV module manufacturing, packaging, and assembly. In 2006, the Solar America Initiative began. Some YDR contracts were completed; however, many YDR contracts were transformed into the new Technology Pathway Partnership project or discontinued before contract completion.

Under PVMaT, direct module manufacturing costs and total capacity among participants were collected annually to analyze the effects of PVMaT and monitor progress. Direct costs of module manufacturing fell from $6.00/Wp in 1992 to $2.92/Wp in 2005 (2008$) (NREL, 2009c). During the same period, capacity increased 18.5 times to 251 MW (Friedman et al., 2005).

Total public expenditures for PVMaT were estimated to be $200.7 million (see Table 4.5). These expenditure data were assembled by reviewing project histories, annualizing by period of performance, and netting out project cancellations and funding adjustments. Program administration expenses were estimated to be 12 per cent of DOE cost-share amounts (Hulstrom, 2010).

Thin-Film PV Partnerships (TFP)

The TFP ran from 1994 to 2008, although PV companies began receiving significant funding for thin-film technology development beginning in 1988. FSA focused on c-Si technologies but had identified two research pathways that were deemed to have the potential to offer low-cost terrestrial PV technology: a-Si and polycrystalline thin films. Through the 1980s and early on in the 1990s, NREL sponsored research that aimed to increase efficiency and reduce instability in a-Si devices. For polycrystalline thin films, NREL sponsored the Polycrystalline Thin Films Subcontract program, which supported the development of CIS and CdTe. In 1994, the a-Si and polycrystalline thin-films research programs were merged to form TFP.

When TFP was launched, c-Si was by far the leading PV technology. However, c-Si cells required large amounts of refined silicon material, and these cells' efficiency was limited by an imperfect band gap. Thin films provided an alternative that held the possibility of overcoming some of the limitations inherent in c-Si, but a significant amount of research would be required to develop thin films into a viable technology alternative. This R&D constituted an investment with high technical and financial risk that few technology companies or investors were willing to make without outside support. The DOE funded nearly all of the materials

Table 4.5 DOE expenditures for PVMaT and TFP

Year	DOE cost share ($ thousands, nominal)			DOE cost share ($ thousands, 2008$)		
	TFP	PVMaT	Deflator	TFP	PVMaT	Total
1988	8900	–	0.62	14413	–	14413
1989	11400	–	0.64	17790	–	17790
1990	10600	–	0.67	15927	–	15927
1991	7800	1179	0.69	11318	1711	13029
1992	8700	7426	0.71	12332	10527	22859
1993	9200	12777	0.72	12759	17720	30479
1994	11000	13412	0.74	14940	18217	33157
1995	12400	11904	0.75	16498	15838	32336
1996	10200	11983	0.77	13318	15645	28963
1997	11480	9825	0.78	14729	12606	27334
1998	16000	9108	0.79	20298	11555	31854
1999	14958	12930	0.80	18701	16166	34868
2000	13205	10421	0.82	16160	12753	28913
2001	18958	8416	0.84	22687	10072	32760
2002	18278	2370	0.85	21525	2791	24316
2003	12495	11457	0.87	14405	13209	27613
2004	10461	8131	0.89	11727	9115	20843
2005	9086	6088	0.92	9857	6605	16461
2006	6134	8424	0.95	6444	8850	15295
2007	6068	9259	0.98	6198	9457	15655
2008	2266	7844	1.00	2266	7844	10110
Total	229589	162957		294292	200681	494973

Note: Dollar values include DOE program administration expenditures.

Sources: Hulstrom (2010); Mitchell (2009); Ullal (2009). GDP Implicit Price Deflator (2005=100) from DoC (2009).

characterization work for thin films, and all interviewees stated that thin-film companies were heavily reliant on TFP and its predecessing initiatives for funding. The DOE's goal was to further encourage development of thin-film technologies and move laboratory research to pilot production. As with PVMaT, cost sharing was an important aspect of TFP, although cost-sharing levels in TFP were lower to reflect earlier-stage R&D.

Total DOE expenditures for TFP were estimated to be $294.3 million (Table 4.5). In general, company cost shares were approximately 30 per cent to 33 per cent of DOE expenses (Hulstrom, 2010).[6] In addition, PV companies reported receiving extensive technical assistance and measurement, characterization, and reliability support from SNL and NREL beyond that provided directly by PVMaT and TFP managers and technical officers. The data presented included funding for centers of excellence and universities that contributed research essential to technology commercialization by thin-film PV companies. Interviewees stated that one of the most significant sources of benefit under TFP was the research and knowledge exchange between university researchers and commercializing companies. Thus, full program funding, not just funding for commercialization partners, was included in the net benefits calculations.

A national research team was formed for each critical area of R&D focus: a-Si photovoltaics; CdTe photovoltaics; CIS photovoltaics; environmental, safety, and health; and thin-film module reliability. Each team generally included about 40 people and was formed from a combination of university researchers, manufacturers, and NREL scientists (Zweibel, 2001):

- Technology partners were major US companies attempting to make the transition to large-scale thin-film manufacturing, and they were allowed up to $1 million/year for a three-year contract. Cost sharing was tiered based on firm size: 40 per cent for large companies and 20 per cent for small companies.
- R&D partners consisted of universities and small businesses that provided basic research support for technology partners. R&D partners were required to cost share at a lower level than technology partners: 20 per cent for large companies and 10 per cent for small companies.

Team participation was a requirement for companies receiving contracts. A percentage of awarded contracts funds was dedicated to team research, while the remaining portion went to proprietary research. Teams helped universities, companies, and national laboratories stay up to date on technological issues, share knowledge, and reduce duplication of research. Core intellectual property development was retained by companies and university partners to incentivize commercialization of the technologies developed.

TFP funded hundreds of subcontracts for more than a hundred different companies and universities. NREL shares seven R&D 100 Awards, awards given annually by R&D magazine, with industry partners for work done under TFP (NREL, 2009a):

- 1984: Boeing, for the first very thin films of a viable material (CIS).
- 1991: Golden Photon, for the first large-area CdTe devices.
- 1998: Uni-Solar, for flexible, waterproof PV roof shingles using triple-junction a-Si.

- 1999: Siemens Solar, for the first large-area CIS modules.
- 2002: BP Solar, for a semitransparent module that can be used in place of glass.
- 2003: First Solar, for the world's first polycrystalline thin-film mass production method, a high-rate module deposition process that can produce one CdTe module per minute.
- 2004: Global Solar, for lightweight, flexible CIS modules that can be easily folded and carried.

Thin films have advanced dramatically over the past two decades, going from about 4 per cent of all US PV production in 1995 to over 60 per cent in 2008. The steep production increase since 2005 is largely due to the success of CdTe at First Solar, the largest US PV producer and a major recipient of DOE funding for CdTe technology R&D.

Because TFP centered on earlier-stage R&D than PVMaT, research had a much broader focus. During the early years of TFP, many contracts focused on exploring different thin-film materials and eliminating those that proved to be unsuitable for PV. Although the benefits of these contracts are more difficult to quantify than for the PVMaT contracts, which often directly reduced manufacturing costs, they nonetheless played a valuable role in accelerating the development of the thin-film industry by guiding companies to the best technological options.

Uni-Solar and BP Solar both brought their multijunction a-Si modules to production through support from the partnership. Although BP Solar has since discontinued its a-Si production line, Uni-Solar is now the second largest US producer and is the largest producer of a-Si. CIS laboratory cell efficiencies increased drastically under TFP (Margolis, 2002). The leading US producer of CIS, Global Solar, also received support from TFP.

Thin-film production by US companies surpassed c-Si production in 2007 (Maycock, 1986–2004; *PV News*, 2005–2009). Figure 4.1 shows the largest US PV producers with cumulative production from 1976 to 2008. Total US PV production reached 1,023 MW in 2008, up more than 50 per cent from just 452 MW in 2007. US production of thin film surpassed US production of c-Si in 2007, largely due to the success of First Solar, which produces thin-film modules with a CdTe semiconductor.

GEOTHERMAL TECHNOLOGIES

In four reports issued by GTP under the main title 'A History of Geothermal Energy Research and Development in the United States, 1976–2006,' GTP identifies four main areas of geothermal technology

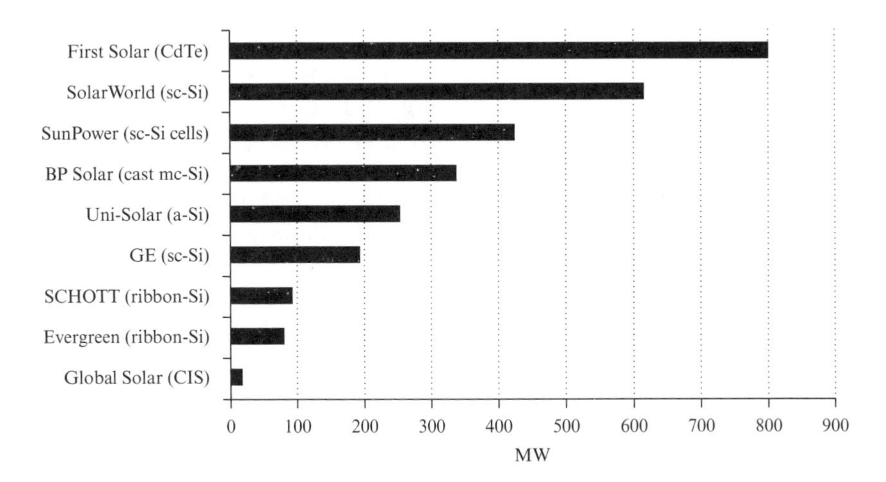

Sources: Maycock (1986–2004); PV News (2005–2009).

Figure 4.1 Cumulative production by US module producers, 1976–2008

development and research: drilling, exploration, reservoir engineering, and energy conversion (DOE, 2010c). This section discusses the history of the GTP program and then describes each of the four main areas of its research. This is followed by a more detailed description of the subset of geothermal technologies selected for analysis in this study.

Background on the DOE's GTP

Limited commercial geothermal electric power production in the United States began in the 1960s. However, following the energy crisis in the 1970s, the development of geothermal resources in the United States became a national priority, and federal and state resources were made available to support R&D and promote implementation projects. As a result, the growth in installed capacity through the 1980s and 1990s was in large part driven by political and financial support.

In the early 1970s, federally sponsored geothermal R&D began with funding from the Atomic Energy Commission (AEC) and the NSF. This was followed by the Geothermal Energy Research, Development, and Demonstration (RD&D) Act of 1974. In 1977, the DOE assumed responsibility for federal geothermal R&D and shortly thereafter created the GTP. The GTPs mission was to support the development of technologies that would improve the economics of tapping less-than-ideal geothermal resources. Prior to 1974, the majority of research on geothermal technology was conducted by government agencies, including the NSF, AEC, US Geological Survey (USGS) and ERDA . From 1974, the RD&D Act instituted the Geothermal Loan Guaranty Program, which provides investment security to the public and private sectors to exploit geothermal resources (EERE/GTP, 2010). ERDA was formed in 1975, and its Division of Geothermal Energy took over the RD&D program. When the DOE was formed in 1977, it took over as the leading agency in geothermal technology research (EERE/GTP, 2008a).

This section discusses the history of the GTP program, starting with a description of each of the four main areas of its research.

Drilling

ERDA, and later the DOE, funded drilling R&D as part of government support for geothermal research in the US, with some costs shared with industry partners. At the inception of DOE's efforts in the 1970s, DOE program managers were responsible for as many as 20 drilling projects. By the early 1980s, SNL assumed responsibility for DOE's drilling technologies program, with some of the work being performed at Los Alamos National Laboratory (LANL) (EERE/GTP, 2008a).

The cost of completing and drilling wells is a major component of the capital investment in a geothermal power plant for both production and reinjection. Research to reduce this cost has been underway since 1975. The primary focus of the DOE research has been to pursue two goals (EERE/GTP, 2008a):

- Develop technologies to lower geothermal drilling costs in the near term.
- Pursue high-risk, long-term R&D activities on advanced concepts that would lead to significant long-run reductions in drilling costs.

Exploration

The DOE's exploration research was initiated by several national laboratories, beginning in the 1970s. Initially, research was conducted by universities and contractors. Since the 1980s, most of the work has been conducted by the University of Utah Earth and Geoscience Laboratory, Lawrence Berkeley National Laboratory (LBNL) and Lawrence Livermore National Laboratory (LLNL). Additional support was provided by Idaho National Laboratory (INL), Oak Ridge National Laboratory (ORNL), LANL and SNL (EERE/GTP, 2008b).

The DOE found that the most efficient way to promote the development of geothermal resources was to have a strong working relationship with the private sector. All of the research was driven by the industry's need to mitigate the highest-risk and highest-cost elements of geothermal resource development. Research programs and projects have been selected based on the projected impact on program goals, especially related to cost of power. To this end, lowering well-field costs through dry-hole avoidance and improving drilling technology were identified as priorities in the increased development of the available hydrothermal resource base. Thus, most of the research work in the exploration area has been focused on these factors (EERE/GTP, 2008b).

Reservoir Engineering

The research projects related to reservoir engineering began in 1976 and were carried out by a variety of institutions, including the national laboratories, universities, the USGS, and the private sector. The DOE's work in this field was focused on three general areas (EERE/GTP, 2008c):

- Improvement of existing technologies to enhance operation and management of geothermal resources by predicting resource productive capacity and longevity (such as reservoir simulation, tracer development and interpretation, reservoir monitoring, and establishing physical and chemical properties of reservoirs and reservoir fluids).
- R&D of innovative technologies for heat extraction (hot dry rock, enhanced geothermal systems).
- Site-specific cooperative studies between American and international researchers to improve reservoir productivity (such as theoretical, modeling, laboratory, and field activities related to demonstration and verification of geothermal resources).

Energy Conversion

When the DOE energy conversion R&D program first began, commercial power production from geothermal resources was limited to The Geysers, a dry-steam plant located in northern California. There was increasing interest in developing geothermal resources; however, since vapor-dominated resources (like The Geysers) are rare, developing the technologies to improve the economic feasibility of using liquid-dominated resources for power production became a primary focus of DOE research (EERE/GTP, 2008d).

A wide range of activities related to energy conversion was conducted in the early research period, with primary emphasis placed on understanding geothermal fluid chemistry and developing materials and components. Geothermal fluids produced from liquid-dominated resources are hot and may contain significant levels of dissolved solids with a higher potential for corrosion and scaling. Thus, the DOE's research focused on identifying compatible materials and minimizing the precipitation of dissolved solids, because these factors determine the feasibility of using liquid-dominated resources for power production (EERE/GTP, 2008d).

Technologies Selected for this Analysis

The GTP has made significant contributions to a wide range of technologies, enabling more effective operation and management of underground resources. This study selected four technologies that had prominent benefits in the geothermal industry and beyond:[7]

Polycrystalline diamond compact (PDC) drill bits

Geothermal systems often require penetrating harder rock than the rock encountered when drilling oil and gas wells, which necessitated the development of improved drill bits. PDC drill bits, with their harder and longer-lasting cutting surface, improved on existing drill-bit technology, allowing the return to a simpler mechanical action from more complex roller-cone action and increasing both productivity (feet drilled per hour) and efficiency (number of drill bits per hour). In the absence of the development of PDC drill bits, it is assumed (based on the interviews with experts) that industry would have continued to use the existing roller-bit technology. Roller bits were an established technology and continue to be used where economically feasible (and where PDC drill bits fail). This failure occurs in hard and fractured rock, formations inter-bedded with stringers, and formations with hard inclusions such as chert.

Binary cycle power plant technology

Binary cycle power plants are an improvement over existing geothermal plant technology and allow the construction of a geothermal plant in sites with lower temperatures that were previously unsuitable for geothermal generation. This technology enabled the development of geothermal plants using low heat sources, thus increasing geothermal capacity in place and offsetting electricity production that uses traditional base load (primarily coal) sources.

TOUGH[8] series of reservoir models

The TOUGH series of models is a new technology representing modeling capabilities not previously available. Reservoir modeling is mainly used as an operating optimization process and, to a lesser extent, during plant design. The benefits of the TOUGH series of models of geothermal applications are reduced drilling costs and decreased uncertainty associated with well management. Because of the flexibility of the TOUGH models, they have also been used for nuclear waste storage, carbon capture and storage applications, and groundwater protection and remediation design of subsurface contamination.

High-temperature geothermal well cements

High-temperature geothermal well cements offer an improvement over existing cement technology. They have a life expectancy of up to 20 years, eliminating annual reworks of geothermal wells. High-temperature cements have also been used in carbon dioxide (CO_2) injection wells on enhanced oil-recovery projects and in capping retired offshore oil and gas wells. The next best alternative would have been to use existing cements (Portland cement).

In summary, the four technologies selected for analysis reflect the wide range of research activities conducted by the GTP and, as a group, have generated significant economic and environmental benefits. Some highlights are:

1. Polycrystalline diamond compact (PDC) drill bits: Approximately 60 per cent of worldwide oil and gas well footage in 2006 was drilled using PDC drill bits (Blankenship, 2009). The main advantage of PDC drill bits over conventional roller-cone bits is that they reduce the frequency of pulling the drill string to replace the drill bit, allowing higher penetration rates and thus reducing the time (and cost) of renting expensive drill rigs. The use of PDC drill bits in offshore applications in the oil and gas industry is estimated to reduce costs by $59 per foot drilled.
2. Binary cycle: In reservoirs where the temperature range is 150°C to 190°C, flash cycle technology is economically viable but has approximately 15 per cent lower electricity generation productivity compared with binary cycle, because of its lower conversion efficiency. Thus, in this temperature range, the next best alternative is a traditional, but less productive, flash cycle geothermal plant.
3. TOUGH models: Using reservoir modeling increased productivity of geothermal resources by an estimated 10 per cent. These benefits are somewhat offset by additional exploration costs associated with reservoir modeling. However, in the aggregate, reservoir modeling has been profitable for the geothermal industry by improving subsurface exploration.
4. High-temperature cement: The rapid deterioration of Portland cement in geothermal wells ($<$12 months) resulted in frequent well workovers and costly well remediation. The use of high-temperature cements enhances performance in terms of structural stability and corrosion

resistance and is estimated to eliminate $150,000 in annual well-remediation costs, as well as extend the working life of geothermal production wells to 20 years or more.

Chapter 6 provides a summary of the findings from a benefit-cost analysis of the DOE's investments in geothermal technologies. For one of the selected technologies, PDC drill bits, the chapter also provides the underlying technology description, primary and secondary data, and calculation underlying the benefit-cost analysis.

VEHICLE COMBUSTION ENGINE TECHNOLOGIES

Early History of Advanced Combustion Research

In 1956, Sandia Corporation established a research branch in Livermore, California (once referred to as Sandia/Livermore and now referred to as Sandia/California). Its early projects focused on advancements in nuclear weaponry, and its programs were closely coordinated with LLNL. This focus continued throughout the cold war.

Because of budgetary cutbacks in the early 1970s and the emergence of energy security as a national priority, SNL in Albuquerque, New Mexico, had diversified into broader areas of energy research. In late 1972, SNL received a research grant from the NSF to conduct a feasibility study related to harnessing solar energy, and Sandia/Livermore became involved in the project. As a result of the oil embargo and the energy crisis in 1973, and with the establishment of ERDA, Sandia/Livermore gained responsibilities in the area of combustion research.

During this period, interest in creating a national combustion research center grew, championed by several scientists at Sandia/Livermore. In 1975, the Combustion Research Program within ERDA was established. It was recognized within ERDA at that time that 'a major shift in national combustion research was necessary, not just a single new project at the principal-investigator level' (Carlisle et al., 2002, pp. 5–6). The purpose of the program was to help industry design and implement new technologies by experimentally validating computer modeling and simulations. Early on, the Combustion Research Program developed links with industrial firms that built and used combustion devices. These partners included General Motors, Ford, Chrysler, Cummins, Caterpillar, Babcock and Wilcox, Combustion Engineering, Bechtel, General Electric, and Westinghouse. Other partners included, from time to time, Exxon, Unocal, and Chevron (Gunn, 2009).

The idea for a research center was based on the belief that combustion research in general had been hampered by the lack of detailed information about the combustion process, and to gain such information, state-of-the-art tools would be needed as enabling technologies. Such tools were very expensive and neither principal investigators nor individual engine companies could justify the expense. A site for the collaborative development and use of such technologies would be needed. In October 1975, ERDA agreed in principle with the concept of a Combustion Research Facility (CRF), and it appeared in President Carter's FY1978 budget at $9.4 million (Carlisle et al., 2002). The DOE Office of Basic Energy Sciences established the CRF and provided the discoveries enabling the technologies used in the design of today's modern internal combustion engines (Eberhardt, 2010).

At the same time, the Office of Energy Research at the former AEC began to focus on fundamental research in chemical sciences with application to energy conversion processes, principally combustion sciences. At the CRF, the Office of Energy Research had responsibility for basic research and for building, equipping and operating the CRF as a DOE-designated User Facility. The Office of Energy Research cooperated with the DOE Conservation Office and the Office of Fossil Energy, encouraging their program to use the CRF's state-of-the-art capabilities. Thus, from early on, a spirit of cooperation, especially among DOE's conservation and basic sciences programs, enabled CRF's work to focus on the often-elusive gap between basic research (usually carried out in scientific laboratories) and applied research (conducted in industry with full-scale devices).

Laser and Optical Diagnostics

> The Combustion Research Facility . . . [was] created with the intent of developing the most advanced diagnostic systems possible for combustion applications, with a special emphasis on combustion in engines. The formula for the program was [. . .] half of the research dedicated to fundamentals in diagnostics and combustion, half of the research dedicated to applications of those tools to problems in practical combustors. (Hartley and Dyer, 1985, p. 27)

Over time, scientists have developed a broad range of spectroscopic methods to probe the electronic structures of atoms and the vibrational and rotational structure of molecules by observing their interaction with electromagnetic radiation. Through various forms of spectroscopy, researchers have been able to identify the chemical species present at different stages of combustion. Laser Raman spectroscopy (LRS) was one such early tool.

LRS was an important early success of laser and optical diagnostics.

In LRS, a monochromatic light source (e.g., a laser) is used to probe a sample, and a detector measures the spectrum of frequencies contained in the light scattered in all directions from the sample. Molecules in the sample may either absorb radiation or contribute to the energy of scattered photons, resulting in a series of output frequencies that provide information about the molecules present.

The advent of LRS allowed the application of Raman spectroscopy to new types of experiments, including many related to combustion research. LRS is one enabling technology for the Advanced Combustion Engine (ACE) R&D subprogram within EERE's VTP, and the CRF led the way in optimizing this tool for visualizing the combustion process:

> [Sandia/Livermore] had developed a new optical capability, but it had never been optimized for use in combustion research. The approach, which used Raman spectroscopy, had been developed at Sandia to look at mixing processes in weapons components. The laser would be focused at a flame and then inelastically scattered off the flame gases. An analysis of the scattered beam would reveal the unstable momentary products of combustion that were released in a particular flame, giving clues about what happened at the flame front. (Carlisle et al., 2002, p. 6)

Also important for laser and optical diagnostics is laser Doppler velocimetry (LDV), a technique by which the direction and speed of fluids (or other materials) can be measured. Particle image velocimetry (PIV) is another diagnostic technique for measuring instantaneous velocity. PIV produces a two-dimensional vector field, while LDV measures velocity at a point.

Mie scattering is an elastic scattering mechanism that occurs when light scatters off of particles with diameters on the same scale as the wavelength of light. The Mie scattering diagnostic is typically applied to particles in the 0.1- to 10-micron range (Asanuma, 1996). Fuel droplets exhibit Mie scattering when probed by lasers, and the scattered light can be collected by a detector to provide information about the spatial distribution of the droplets. Mie scattering is a useful phenomenon in a variety of combustion experiments, including those that focus on airflow and fuel spray. In-cylinder airflow can be observed and quantified in real time by scattering light off of particulates introduced into the airflow stream. Similarly, the distribution and evaporation of fuel droplets can be observed during diesel injection experiments. Information on spatial and temporal distribution is particularly useful for understanding and improving the dynamics of fuel injection.

Rayleigh scattering is similar to Mie scattering, but it occurs with smaller particles and atoms or molecules in the gas phase. Although Mie scattering occurs when the particle diameter is similar to the wavelength of incident light, Rayleigh scattering occurs when the particle diameter is much smaller than the wavelength of light.

A variety of spectroscopic techniques are used to probe the combustion process in the laboratory, and many were applied to engine combustion in the early 1980s. Laser-induced fluorescence (LIF) and tracer-based LIF are diagnostic tools that allow for the observation of light species such as the molecule OH and various molecular species that are common in combustion. Light species are particularly difficult to interrogate using other spectroscopic methods because very high-energy (ultraviolet) sources are required for optical excitation. The energy difference between the ground and excited states of light species can be on the order of 10 electron volts (Forch et al., 1990). In LIF, a fixed-wavelength or tunable laser is used to interrogate species in a combustion chamber. These species emit lower energy wavelengths that provide information about the vibrational–rotational states of molecules. The emission spectrum from the molecules of interest is sometimes complicated by interference from other emission processes, such as combustion luminosity.

Laser-induced incandescence (LII) is the emission of radiation that occurs when a laser beam interacts with soot or other particulate matter (AIAA, 2009). This technique can be used in the laboratory to determine information about the average properties of soot that forms as a combustion product. The temperature of particulate matter rises when it absorbs incident laser light, and the heat generated is then emitted as thermal radiation. At very high temperatures, the soot or other particulate matter may vaporize. Like all laboratory techniques, LII has limitations, including complications with high soot loadings and long path lengths. In either case, signal attenuation is likely (AIAA, 2009).

Improvements to semiconductor diode lasers that operate at room temperature in the visible and near-infrared areas of the spectrum have contributed to advances in the ways in which laser absorption spectrometry (LAS) is applied to combustion research (Allen, 1998). LAS is based on the principle that different molecular species absorb light of different wavelengths. New laser diodes have expanded the range of species that can be monitored using LAS; for example, lasers that emit in the infrared region have enabled better detection of species, such as carbon monoxide, that absorb infrared wavelengths (Hanson et al., 2002). Improvements to sensor technologies that detect and identify the species present in a sample have also furthered the usefulness of LAS to combustion analysis. Because real-time monitoring is possible using LAS, the technique is employed to analyze engine combustion gas flows (Mattison et al., 2007).

Combustion Modeling

Over the years, a number of national laboratories and universities have been involved with the Combustion Research Program within the CRF for specific research purposes: LLNL in the area of combustion chemistry; LBNL and University of California–Berkeley in the area of homogeneous charge engines and processes; LANL in the area of large-scale computer models; Purdue University in the area of heat and mass transfer; Princeton University in the area of direct fuel injection engines; the University of Wisconsin in the area of experimental engineering processes; the MIT in the area of flame propagation. Fluid mechanics in engines was studied at both Pennsylvania State University and Imperial College (London).

Combustion modeling allows researchers to conduct so-called experiments much more quickly than they could in the laboratory. Such modeling has thus expedited the discovery of new combustion engine technologies. It is important to note, however, that modeling results are only helpful when verified by a subset of empirical data. It is the combination of advanced spectroscopic techniques and increasing computational capability that has provided a basis for innovation with respect to advanced combustion engines in the VTP (Eberhardt, 2009).

In 1982, LANL developed the so-called KIVA codes, which simulate the fluid dynamics of combustion processes in internal combustion engines.[9] However, computers at that time were not fast enough to make the tool practical. In 1983, LANL began working with a small community of potential adopters of the KIVA technology and shared their codes with General Motors, Cummins Engine, and others. The software was released to the public in 1985, and throughout the development of the KIVA codes, government scientists worked closely with industrial partners and others in the user community (Amsden and Amsden, 1993). In fact, the CRF had dedicated laboratory space for visiting researchers from partner automotive companies who spent months at a time contributing to the project (Eberhardt, 2009).

Adoption of the latest version of the KIVA codes is widespread, and users include Caterpillar, Cummins Engine, General Motors, Ford, and Chrysler (Amsden and Amsden, 1993). Some of the patents for vehicle technologies in the automotive industry specifically cite the KIVA codes as an enabler of the inventions. In addition, the code has broad applications beyond modeling combustion in vehicle engines, and it has been used for modeling gas turbines, incinerators, and waste heaters.

From an economic perspective, the KIVA codes are similar to a general-purpose technology (GPT) in that they leverage the application of laser and optical diagnostics. A GPT has the characteristics of pervasiveness, an inherent potential for technological improvements, and innovational complexities that give rise to increasing returns to scale in R&D (Bresnahan and Trajtenberg, 1995).

Direct Injection Diesel Engine

Diesel engines are a type of combustion engine in which fuel ignites when compressed. Diesel engines are traditionally known for high efficiency but also for emitting high levels of oxides of nitrogen (NO_x) and particulate matter (PM). Modern diesel engines burn cleaner than their traditional counterparts as a result of advances in producing cleaner diesel fuels, modification of the air handling and combustion system resulting from improved understanding of diesel combustion, introduction of electronic control of engine functions, major improvements in fuel injection equipment, and employment of various emission control techniques such as exhaust gas recirculation.

In the first stroke of a four-stroke combustion ignition engine, the piston moves away from the intake valve, drawing air into a cylinder. Next, the piston compresses the air, and fuel is injected at the end of this second stroke (when air is at maximum pressure and temperature). As ignition and combustion occurs, the piston is forced downward by the expanding gases, after which the piston swings upward on the fourth stroke, clearing post-combustion gases from the cylinder. The high efficiency of diesel engines is the result of high compression ratios, rapid combustion, and the ability to control engine load through the quantity of fuel injected (as opposed to controlling load by restricting the intake airflow, as is used in spark-ignition engines). The high temperatures associated with both the high compression ratios and the ignition properties of diesel fuel enable the fuel/air mixture to spontaneously ignite.

Although the basic design of the engine has not significantly changed over the past century, today's direct injection diesel engines have much lower emissions than the previous generation of indirect injection diesel engines. In older diesel engines, fuel and air were mixed in a precombustion chamber prior to injection into the cylinder. Because the mixing and injection steps were mechanically controlled, they could not always be optimized for specific engine conditions and often led to the release of uncombusted fuels. Modern, direct injection equipment is computer controlled and designed to deliver the optimal amount of fuel at the optimal time. Compared with older diesel engines, today's direct injection diesel engines are characterized by higher efficiency (fuel economy), lower emissions (for some categories of pollutants), and higher power.

The major advantage of direct injection is increased efficiency, not reduced emissions. Early direct injection diesel engines injected the fuel when the piston was top dead center, and fuel burned very efficiently. However, at such peak flame temperatures, there are high emissions of NO_x and PM. Emissions of these two classes of pollutants have been reduced by over 90 per cent in modern direct injection diesel engines, but they still create a challenge in meeting US emissions standards (DOE, 2008b). NO_x and PM emissions can be reduced with the introduction of an aftertreatment technology, but the high cost of such technologies must be compared with other potential technologies for in-cylinder reductions of emissions before this technology is widely adopted (Nam, 2004).

Advances in engine technologies themselves have also occurred. These include high-pressure electronically controlled fuel injection, exhaust gas recirculation (EGR), and new combustion approaches such as homogeneous charge compression ignition (HCCI).

High-pressure electronically controlled fuel injection for heavy-duty diesel engines is a system in which standing high pressure exists in a common rail. The system uses fuel injectors that are controlled electronically to deliver fuel appropriate to changing engine demands and to optimize performance (Fort et al., 1980).

EGR is a technology used in both gasoline and diesel engines to reduce NO_x emissions by lowering the temperature in the combustion chamber. EGR involves recirculating a fraction of the exhaust back into the intake stream. In diesel engines, the exhaust replaces oxygen in the precombustion gas mixture. The exhaust is first cooled and compressed, allowing a larger volume of gas to be reinjected. Although EGR reduces NO_x emissions, it may also increase PM emissions. The goal is to reduce engine-out particulate emissions to reduce the demand on (or need for) aftertreatment (i.e., diesel particulate filters). Materials engineering may provide solutions for mitigating durability issues currently associated with EGR. Because exhaust chemistry varies with choice of fuel, EGR must also be optimized for different operating conditions (Lance and Sluder, 2009).

HCCI refers to a strategy employing chemical-kinetically controlled volumetric combustion of a mostly premixed charge. The strategy differs from conventional combustion-ignition engines by avoiding the rich burn during fuel injection and from conventional spark-injection engines by avoiding flame propagation. As a result of this process, fuel efficiency increases and NO_x and PM emissions decrease.

The application of laser and optical diagnostics and combustion modeling has made significant contributions to the study and development of improved diesel combustion, EGR, and HCCI strategies. New fuel-injector technologies came out of advancements in electronics, but visualization diagnostics have provided important details for optimization of spray targeting, evaporation, mixing, ignition, and combustion.

DOE-funded research performed using laser and optical diagnostics and combustion modeling has contributed to spark-injection engine improvements starting in the late 1970s, and research applicable to diesel engines began in the mid 1980s and continues today within EERE (Siebers, 2009). The major impact of this research on heavy-duty diesel engines began in the mid 1990s, and it too continues today. Resulting enhancement of the understanding of in-cylinder processes has contributed to improvements in both engine thermal efficiency and engine-out emissions.

In an effort to promote efficient combustion without reaching the peak flame temperatures and without reducing performance (i.e., fuel efficiency), the DOE funded research on injection rate shaping (i.e., the way in which the fuel is injected) during the 1980s. The goal of the technology is to shape the pressure rise by controlling the rate at which the fuel is injected. With fuel being injected optimally throughout the stroke process, more fuel reacts with oxygen and less nitrogen combines with oxygen to produce harmful NO_x emissions. By optimizing fuel injection, emissions are reduced without compromising engine performance.

In summary, improving engine efficiency and simultaneously meeting stringent new emissions regulations required significant new and detailed knowledge of diesel combustion processes. Without this understanding, engine designers would have been left with decades of cut-and-try approaches to arrive at the required improvements in engine design. Laser diagnostics and optical engine technologies allowed the combustion process in an operating diesel to be probed and measured in real time. The understanding developed with laser diagnostics and other optical methods has had two impacts. First, it directly helped engine designers improve diesel designs by providing an accurate picture of how diesel combustion occurs and scales with a multitude of engine parameters. Second, the data and understanding allowed the validation of computer models for predicting diesel combustion. These models are now widely used by automotive and engine companies to design and optimize diesel engines. Together, these two impacts have led to greatly improved diesel engine designs and efficiency.

CONCLUSIONS

The three case studies provide a snapshot of how EERE engages in technology development through the life cycle of energy technologies. The key point is that in addition to the importance of aggressively pursuing R&D that will enable the next generation's energy infrastructure is the necessity to pursue near-term solutions and enhancements to existing and mature energy technologies. Long-term energy needs must be balanced with short-term energy needs.

The following three case studies document the economic impact from selected EERE efforts, focusing on applied R&D (Solar Energy Technology Program, SETP), market technology (Geothermal Technologies Program, GTP), and improvements in mature technology (Vehicle Technologies Program, VTP). In all three cases, significant energy savings and social return were realized from taxpayers' investment.

NOTES

1. During part of the period covered in this study, the National Center for Photovoltaics (NCPV) coordinated DOE's strategy for photovoltaics. Solar projects are conducted by DOE, its national laboratories (particularly NREL in Golden, Colorado, and SNL in Albuquerque, New Mexico), university research centers, nonprofit centers of excellence, and solar energy technology companies. In addition to NREL and SNL, other key participants in the NCPV at present are Brookhaven National Laboratory, the Georgia Institute of Technology, the Institute for Energy Conversion at the University of Delaware, DOE's Southeast Regional Experiment Station, and DOE's Southwest Technology Development Institute.
2. In 2000, PVMaT was renamed the PV Manufacturing R&D Project to reflect changes in PV manufacturing technology needs; however, it was still commonly referred to as PVMaT, which is the name used in this report for simplicity.
3. PVMaT originally focused on module manufacturing before expanding to include balance of system (BOS) components and system integration elements, but these accounted for less than 15 per cent of total DOE funding for PVMaT.
4. Cost-sharing levels differed by project phase. Overall industry cost sharing for all phases was about 48 per cent. To encourage collaboration with universities, companies were allowed to waive the cost-sharing requirement up to a specified amount for contract funding used to conduct research at universities. Smaller companies were required to meet a lower cost-share percentage than larger companies.
5. Panels generally included one representative each from NREL, DOE, and SNL, plus ten or more other experts with varied PV experience and no conflict of interest (Margolis, 2002).
6. Though annual TFP cost data from annual reports were only available for 2004 through 2008, data for 1988 through 2003 were obtained from Photovoltaics Energy Systems annual reports, NREL annual reports, or budget justification documents. Sources for TFP were Smolter and Stuart (1996) for 1988 to 1995; Office of Energy Research (1997–1999) for 1996 to 1998; Office of Science (2000–2004) for 1999 to 2004, and EERE (2005–2009) for 2004 to 2008.
7. The technologies were selected based on a review of the published literature and the DOE's historical summary reports (EERE/GTP [2008a], EERE/GTP [2008b], EERE/GTP [2008c] and EERE/GTP [2008d] EERE/GTP [2010]). The technologies were intended to capture significant contributions by DOE across the broad range of geothermal research conducted by the GTP.
8. TOUGH is both an acronym for 'transport of unsaturated groundwater and heat' and a reference to tuff formations in Yucca Mountain, which was one of the first major applications of the code.
9. A 'kiva' is a subterranean room used for religious purposes by the Pueblo people of the Los Alamos region (Eberhardt, 2009). The name of the codes reflects the geographic area in which they were developed.

5. Investments in solar energy technologies

This chapter presents the benefit-cost analysis of the DOE's investments in Photovoltaic Energy Systems. As reviewed in Chapter 4, the DOE helped the US PV industry in the development, scale-up, and maturation of core PV technologies and manufacturing processes. Benefits accrued directly to PV module producers and consumers in the form of increases in product quality, operational efficiency, reliability, and reductions in production costs. Following methodologies pioneered by Griliches (1958) and Mansfield et al. (1977), economic benefits were quantified by comparing actual technological progress to counterfactual scenarios under which DOE technical expertise, technology infrastructure, and financial support were not available and PV module companies pursued their technology R&D strategies without DOE support. Our approach was to conduct primary and secondary research on technology advances in photovoltaics funded or co-funded by the DOE and ascertain how, when, or if those advances would have been made in the absence of the DOE's programs. This process defined the next best alternative against which economic benefits were measured and, definitional to this approach, established attribution to the DOE. Where technical accomplishments may have economic impacts outside the PV market, such as the accelerated adoption of wire saw technology in the semiconductor industry or processes for refining high-grade silicon, these externalities were also included in the quantitative analysis.

OVERVIEW OF THE BENEFIT-COST ANALYSIS

Within the Photovoltaic Energy Systems cluster, the technologies of focus were those that supported the development of c-Si and thin-film modules, including R&D for solar cells, module manufacturing, and technology infrastructure. FSA, PVMaT, and TFP were technology initiatives that represented ten-year commitments (at least) on the part of the DOE to provide cost share and technical expertise to US companies seeking to develop novel commercial PV technologies. The core of PV systems are the modules, and given this central role, the extent to which DOE enabled, accelerated, or supported module R&D constituted a research question of keen interest.

A large body of engineering and public policy literature had rigorously assessed technical progress in photovoltaics and the role of the DOE in supporting that progress. Many recent works, including those by Komp (2001), Green (2005, 2009), Swanson (2006), and Osterwald and McMahon (2009), continued to highlight the significance of the 1975–1985 FSA project on the development of solar technologies.

DOE program reviews and technical reports for PVMaT and TFP highlighted the challenges PV companies faced in further developing core PV technology, operational efficiency and economies of scale in manufacturing. These challenges inhibited sector and technology development. NREL, in particular, offered several prospective analyses of program goals and strategies (Mitchell et al., 1992; Surek, 1992; Witt et al., 1993), as well as retrospective assessments and case studies of technical progress and best practices to guide future endeavors (Witt et al., 2001; Margolis, 2002; Margolis et al., 2006).

NREL, SNL, and the DOE jointly performed two internal quantitative analyses of investment recovery on the DOE and private investment in PVMaT (Witt et al., 2001; Friedman et al., 2005).[1] Figure 5.1 presents historical results from these analyses through 2005 and the authors' best forecasts through 2011. Witt et al. (2001) found that the public investment in PVMaT was recouped in 1997, and the industry investment was recouped in 1999. Friedman et al. (2005) calculated that, between 1992 and 2005, the average module manufacturing cost fell 54 per cent, production capacity increased 18.5-fold, and progress ratios were 87 per cent for c-Si companies and 81 per cent for thin-film companies. These results suggested that economic benefits might be significant and should be quantified in the economic benefit-cost analysis.

Primary Data Collection

Primary data were collected via semi-structured interviews to quantify the role the DOE played in furthering PV technology, reviews, and synthesis of technical impact data from the science and engineering literature. Interviews were conducted with representatives of:

- PV companies and other recipients of DOE cost shares
- scientists, engineers, and policy analysts with DOE national laboratories
- academics and university-based researchers

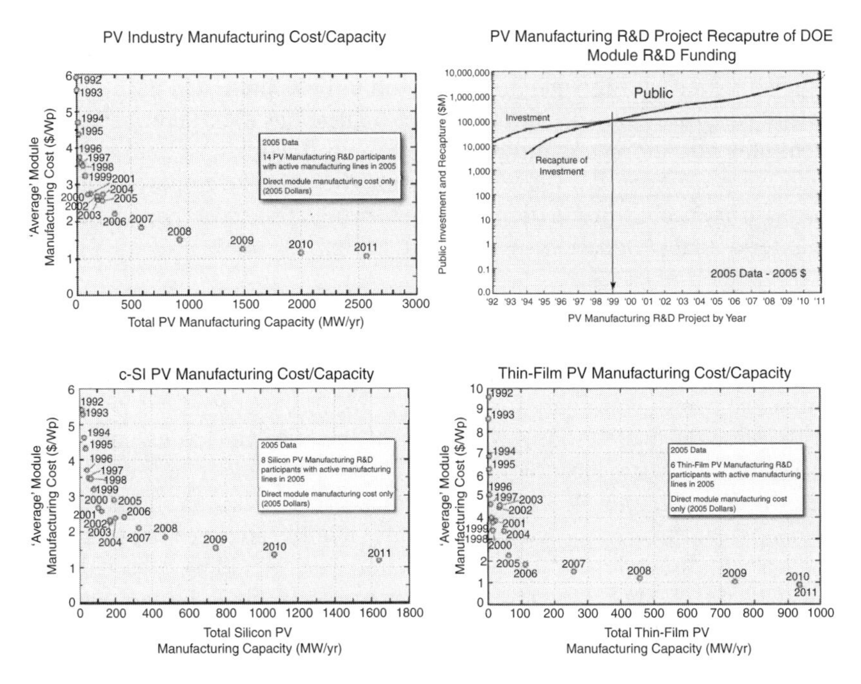

Source: Friedman et al. (2005).

Figure 5.1 Results from 2005 DOE investment recovery analysis

- energy-related trade associations
- retired company executives and government experts from DOE, NASA, and other agencies, which was important for reviewing the state of technology between 1970 and 1985
- venture capital and technology consulting groups
- investor-owned electric utilities.

Discussion topics included technologies developed under FSA, PVMaT, and TFP (see Chapter 4); the role and significance of the DOE and DOE cost sharing; counterfactual technology development and technical progress; US and non-US innovation policies for photovoltaics and technology infrastructure. All interviewees' responses, especially those receiving DOE cost share, were compared with extant technical literature, market analyses, and reviews of non-US programs.

We respected the sensitive nature of the information provided by participants. Candid assessments of technology development were needed to quantify economic benefits and determine attribution to DOE. Interviews were confidential, as were the names and affiliations of private-sector participants. Participants were informed that their comments, as well as any supporting data or documentation, would only be presented in the aggregate. Therefore, firm-specific responses to questions about counterfactual technology development and reductions in production cost per watt over time were confidential. All cost data are weighted averages calculated using firm-specific production, baseline production cost per watt, and counterfactual production cost per watt, data.

Next-Best Technology Alternative

Economic benefits are measured relative to the next-best technology alternative. The next-best alternative for c-Si modules was modules produced in the absence or delayed introduction of the efficiency gains, manufacturing technologies, technology infrastructure, and other innovations as described in Chapter 4. Companies' rates of progress, as measured by year-on-year production cost reductions, would have been lower. For example, in discussing how their production costs would have been different without PVMaT, the most common comment made by interviewees was that PVMaT enabled firms to work on issues of long-term importance.

In a start-up environment, firms often put off long-term goals to focus on near-term ones that are of immediate concern for keeping the business going. Ultimately, these long-term projects are critical to the maturation and sustainability of a company.

The next-best alternative for thin-film modules was a counterfactual c-Si module produced in any given year. A-Si, CdTe, and CIS/CIGS modules would not have emerged as commercial products before 2008. TFP was characterized as 'fundamentally enabling,' and academic and industry researchers alike doubted whether thin films would have been viable without DOE support. One principal scientist posited that '[t]here were too many problems, progress was at times too slow, and it took so much time to get there [commercialization] that it is hard to see how thin films would have made it to the marketplace without DOE.' Where there were, in actuality, labor and materials savings relative to c-Si modules, these would not have accrued. First Solar is often offered by the industry as a success story, but even this financially successful firm relied on TFP funding from its start in 1991, even after it received private equity financing in 1999. Commercial production at First Solar did not commence until 2003, more than 12 years after the company's founding. Thus, both thin-film and c-Si modules were compared against the same alternative: counterfactual c-Si modules.

Technology Acceleration

Studying when technology milestones would have been met in the absence of the DOE is a technology acceleration analysis. Having a more cost-effective process today, rather than tomorrow, offers both a resource saving and time value of money impact. Not only may initiatives lead to innovations that might not have been developed, they also broaden R&D programs, which in turn accelerate the accrual of benefits for society. In the case of PV modules, superior technology performance and lower costs, as well as earlier accrual of these benefits, combined to amplify economic benefits.

Technology acceleration was a critical area of analysis particularly because of the foundational role of FSA. Before the early 1980s, PV modules were vastly inferior to modules that emerged just a few years later during FSA's block purchases. Developing superior modules to meet FSA specifications enabled the PV industry to move from supplying modules for off-grid niche market applications to on-grid residential and utility applications. One interviewee likened the breadth of technology developed during the FSA period to a recipe book for a PV industry: cell, module, and systems technologies, requirements for technology infrastructure, and processing and production automation technologies were all dramatically improved. In the absence of FSA's effect on the terrestrial PV industry, PV modules would likely have improved at a far slower rate along all relevant dimensions.

Because there was no US substitute project for FSA, contemporary programs funded by other national governments were reviewed for a next-best alternative—but none were found. In 1974, following the 1973 oil price shocks, Japan created an energy R&D program called the Sunshine Project, which was intended to support PV, coal gasification and liquefaction, and geothermal and hydrogen technologies. The Agency of Industrial Science and Technology within the Ministry of International Trade and Industry organized the project. Much of the PV funding from the Sunshine Project was directed toward developing a-Si and c-Si, including research on low-cost silicon feedstock material. Japanese a-Si companies had produced a-Si mostly for small applications, such as watches and calculators, before improving cells for large-scale use. With the Sunshine Project, Japan began to pursue the development of grid-connected rooftop PV systems, leveraging technology outcomes and best practices published by JPL and discussed widely in the global PV technical community. Large plots of land needed for array fields are rare and expensive in Japan because of its mountainous topography, and the imperative for the Sunshine Project was to facilitate grid-connected distributed power, which today accounts for the majority of PV applications in Japan (Kurokawa and Ikki, 2001). The New Sunshine Project replaced the Sunshine Project in 1993. The New Sunshine Project shifted from the earlier focus on R&D to emphasize commercialization. Experts believe that although the Sunshine Project was important, it benefited greatly from FSA and, therefore, should not have been a considered a substitute for FSA.

Technical and Economic Impact Metrics

Technical and economic impact metric pairs are used to operationalize economic models that calculate benefits. A technical impact metric conveys the benefit of a new technology in terms of physical units, such as number of labor hours saved or amount of raw materials saved, relative to the next-best alternative. An economic impact metric, such as wage rates or cost of materials per ton, monetizes that technical benefit. The product of technical and economic impact metrics is then applied to the relevant quantity of output to derive total economic benefit.

The data required for this analysis included:

- production cost per watt for each company
- guaranteed PV module reliability measured in years
- annual volume of PV modules produced (in megawatts) for each company
- annual volume of PV modules installed in the United States (in megawatts).

The breadth of technology developed and reviewed in Chapter 4 presented the challenge of how best to collect data to inform technical and economic impact metrics and then aggregate across technologies and companies. The solution was to use the common PV industry progress measure: production cost per watt.

PV companies and DOE's technology and policy strategists all placed great emphasis on driving down the production cost per watt for modules, which accounts for a significant portion of the total installed cost of a PV system. Gains in efficiency, technologies from process development R&D, yield gains, and other technical impact metrics all influence the production cost per watt. This progress measure has been used and commonly reported since the late 1970s. Actual and counterfactual production cost per watt and production quantity data were aggregated across all funded PV companies. To the best of their ability, interviewees isolated technology effects from the addition of new production lines or similar capacity increases.

A second technical metric of interest was reliability, as measured by the guaranteed life of a PV module, which is not captured by production cost per watt. Gains in reliability benefit consumers directly by lowering the annualized module cost and thus the levelized cost of electricity (LCOE). The technical impact metric was lifetime measured in years, and the economic metric was the change in the annualized module cost, which also incorporated decreases in the production cost per watt. The economic results section includes the formula for calculating this benefit.

Actual PV module quantity output was used as the quantity basis for calculating total economic benefits. Using historical production data enabled the capture of economic benefits from technology development for every unit of production and every unit installed. Innovation both increased the megawatt rating of PV modules and, through income and substitution effects, increased module demand. A significant positive attribute of using historical production data is that this analysis did not change the timing of total PV system investments or public costs associated with demand-side policies, thereby isolating economic impacts attributable to technological development only.

Treatment of Demand-side Policies, Rebates, and Financial Incentives

The market for PV systems is global and driven by public policy initiatives (Jennings et al., 2008; Wiser et al., 2009). Feed-in tariffs, renewable energy portfolio standards, tax credits, and rebates create a demand-side pull that accelerates the accumulation of PV installations. Federal, state, local, and foreign governments and authorities' public policies have sustained a market for photovoltaics. In the absence of demand-side policies, and without regard to externalities, grid-connected PV systems would not be cost-competitive with fossil fuel or other energy sources during this study's period of analysis.

German, Spanish, and Japanese policies, as well as those in many other countries, were critical in supporting the push for scale in R&D and manufacturing and accelerating the accumulation of installed PV systems. Germany, which enacted a feed-in tariff in 2004, increased its solar capacity by more than a factor of five by 2008 despite relatively low sunshine levels. Japan, which surpassed US installations in the late 1990s and continued to grow steadily into the 2000s, is now experiencing a decreasing growth rate after a key incentive program was phased out in 2005. In Spain, which in 2007 enacted a feed-in tariff and a building code that requires newly constructed or renovated commercial buildings to generate a portion of their electricity from photovoltaics, PV installations more than quadrupled from 2007 to 2008.

In the US, many state and local governments, nonprofits and utilities also offer incentives for photovoltaics, including loans, rebates and commercial and residential tax credits. Several states also mandate renewable portfolio standards. In California, which had more installed PV capacity in 2008 than

Table 5.1 Sampling of federal incentives for photovoltaics

Year	Initiative
1978	Department of Energy Act of 1978 allocated $13 million for PV systems in federal facilities
	National Energy Conservation Policy Act of 1978 authorized $98 million for the federal Photovoltaic Utilization Program
	Energy Tax Act of 1978 created a 10% business tax credit for photovoltaics
	Public Utility Regulatory Policies Act of 1978 required utilities to purchase from small renewable energy producers
1980	Crude Oil Windfall Profit Tax of 1980 created a 40% residential energy tax credit that could be used for photovoltaics and raised the business tax credit to 15%
1981	Economic Recovery Act of 1981 authorized accelerated depreciation of PV equipment
1986	Tax Reform Act of 1986 reinstated the business credit for photovoltaics at 15%, dropping to 12% in 1987 and 10% in 1988
1988	Technical and Miscellaneous Revenue Act of 1988 extended the PV business credit through 1989
1989	Omnibus Budget Reconciliation Act of 1989 extended the PV business credit through 1990
1990	Omnibus Budget Reconciliation Act of 1990 extended the PV business credit through 1991
1991	Omnibus Budget Reconciliation Act of 1991 extended the PV business credit through 1992
1992	Energy Policy Act of 1992 permanently established a 10% PV business credit and formed the Renewable Energy Production Incentive offering 1.5 cents per kilowatt-hour (kWh)
2005	Energy Policy Act of 2005 raised the business tax credit to 30%

Source: Margolis (2002) and Database of State Incentives for Renewables and Efficiency (NC Solar Center, 2009).

any other state, more than 50 nonfederal financial incentives are available for photovoltaics, compared with fewer than ten in most states. Each of the ten states with the highest PV capacity in 2008 has a renewable portfolio standard. Half of these ten states are among the 18 states that offer both individual and corporate tax credits (NC Solar Center, 2009).

This analysis left US and international, state and local demand-side policies unchanged (Table 5.1), thereby calculating actual accrual of economic benefit of the DOE's technology development additionality for every unit produced and/or installed. Availability of technology likely influenced policy design and funding requirements, but incorporating counterfactual policy analyses would have diluted the analytic focus from valuing the contributions of DOE technology development programs to that of the contributions and roles of all public and private stakeholders in growing the installed base of PV systems overall.

The effect of demand-side policies on measures of economic return is indeterminate. Certainly, international policies have stronger effects than US policies (60 per cent of PV production by US companies is destined for international markets) which encourages economies of scale in manufacturing, thereby lowering production cost per watt. Setting aside international policy considerations, adjusting the timing and introduction of demand-side policies would shift the timing, frequency, and/or occurrence of public and private investment decisions in the PV industry, in non-DOE public-sector subsidies and other outlays, and in the rate of system installation accumulation. In turn, the accrual of economic benefits detailed in this analysis (as well as environmental health, GHG, energy security, and other benefits) would be affected.

Although such a study has great merit, focusing on the technology variables isolated the DOE's impact on technological development, which was the principal objective of this analysis. Subsequent demand-side policy studies can leverage of the results of this study's economic analysis of the DOE's effect on technology development.

Attribution to the DOE

In this study, the challenge posed by attribution was avoided because research questions focused on program additionality and interviewees understood that their responses should reflect such a focus. All counterfactual production cost per watt data (i.e., insights into how those historical cost data would be different) were provided by PV companies under the assumption that DOE technical expertise and cost sharing were not available and companies' progress continued in its absence. Thus, attribution of economic benefit to the DOE was implicit in the approach.

BUDGET HISTORY

Photovoltaic Energy Systems received the largest portion of ERDA's and DOE's budget for solar energy initiatives between 1975 and 2008. Over this period, total appropriations from Congress for solar energy were $4,089 million in nominal terms, or $7,438 million in real terms (2008$). Photovoltaic Energy Systems accounted for over half of these funds: $2,309 million in nominal terms, or $3,710 million in real terms (see Table 5.2).

Table 5.2 DOE investment in photovoltaic energy systems, 1975–2008

Year	Total, Photovoltaic Energy Systems (nominal millions)	Total, Solar Energy Program (nominal millions)	Deflator	Total, Photovoltaic Energy Systems (2008$ millions)	Total, Solar Energy (2008$ millions)
1975	0.60	0.60	0.31	1.94	1.94
1976	21.56	89.21	0.33	65.90	272.69
1977	59.40	248.31	0.35	170.69	713.57
1978	76.20	232.10	0.37	204.61	623.24
1979	118.80	324.10	0.40	294.50	803.44
1980	150.05	378.10	0.44	340.88	858.97
1981	151.60	363.17	0.48	314.91	754.39
1982	74.00	152.05	0.51	144.87	297.68
1983	57.92	118.96	0.53	109.07	224.04
1984	50.18	110.23	0.55	91.09	200.07
1985	54.65	97.73	0.57	96.28	172.18
1986	40.30	74.02	0.58	69.47	127.59
1987	40.25	46.15	0.60	67.43	77.30
1988	34.69	56.90	0.62	56.17	92.14
1989	35.15	52.26	0.64	54.85	81.56
1990	34.33	54.25	0.67	51.58	81.51
1991	46.07	67.09	0.69	66.85	97.35
1992	60.00	90.75	0.71	85.04	128.64
1993	64.90	94.81	0.72	90.00	131.49
1994	74.88	111.05	0.74	101.70	150.83
1995	83.84	118.50	0.75	111.54	157.66
1996	61.27	87.20	0.77	79.99	113.86
1997	59.21	83.41	0.78	75.97	107.02
1998	64.69	83.63	0.79	82.07	106.10
1999	70.56	90.91	0.80	88.22	113.66
2000	64.57	81.41	0.82	79.02	99.63
2001	74.26	91.69	0.84	88.87	109.73
2002	65.46	87.11	0.85	77.09	102.58
2003	73.25	82.33	0.87	84.44	94.91
2004	72.54	80.73	0.89	81.32	90.50
2005	65.84	75.73	0.92	71.43	82.15
2006	32.41	41.14	0.95	34.05	43.22
2007	138.37	157.03	0.98	141.33	160.38
2008	136.74	166.32	1.00	136.74	166.32
Total	2308.52	4088.98		3709.91	7438.33

MODEL FOR ESTIMATING ECONOMIC BENEFITS

This section begins by reviewing baseline data on PV modules and then presents the economic model that compared that baseline data with data on counterfactual technological progress collected from interviews with subject matter experts. The difference between actual and counterfactual production cost per watt and reliability constituted the majority of quantified economic benefits. Benefits were also calculated for technology spillovers into the semiconductor industry, specifically for the UCC silicon refining process and accelerated adoption of wire saw technology. All dollar values in this chapter are in real terms (2008$), unless otherwise specified.

Key Terms and Data Sources

Myriad market research reports offer what are, at times, conflicting values for any given year for four key variables imperative to this analysis: (1) production quantity, (2) production cost per watt, (3) guaranteed module reliability, and (4) PV installations in the United States. Definitions, assumptions, and data sources for these variables, whose values are presented in the tables and figures in the remainder of this case study, are provided in the following discussion.

The following metrics are used extensively throughout the remainder of this section:

- efficiency (specifically, conversion efficiency)
- power
- installed cost per watt
- production cost per watt
- reliability
- LCOE.

For PV cells, efficiency is defined as the ratio of electric power generated by the solar cell to the amount of incident solar power. If a solar cell illuminated by 100 W of solar power generates 15 W of power, the cell's solar energy conversion efficiency is 15 per cent. In this chapter, the terms 'efficiency' or 'efficient' without a modifier always refer to the solar energy conversion efficiency. When referring to manufacturing or costs, this book specifically uses the term 'operational efficiency.'

Power is the rate at which energy is supplied by the PV cell or module. The amount of power contained in the solar spectrum hitting a given area is not uniform across the globe. Therefore, standard test conditions of 1 kW/m^2 at 25 °C were established to allow researchers and companies to communicate performance measurements comparably. Thus, all power ratings for solar cells and modules are reported subject to conditions that have been artificially defined, not what they will experience in the field.

The installed cost per watt of a PV energy system refers to the sum of all modules, balance of systems, installation, and other costs divided by the power rating of the system. This study quantitatively evaluates only the PV module component of the system. The common metric for reviewing manufacturing costs for PV modules is the production cost per watt. Production cost per watt captures increases in conversion efficiency and increases in operational efficiency of production systems. As cell technology improves, so does the cell's efficiency rating, which lowers the material's cost per watt and increases the power rating of a module. Improvements in manufacturing technology also place downward pressure on production cost per watt. The following data sources and estimation procedures were used:

- **1974 to 1989:** Data were estimated by subtracting the estimated gross margin from average price data reported by the Energy Information Administration (EIA) or presented in FSA reports. Any missing years were linearly interpolated using the 1992 EIA average module price as the final data point.
- **1990 to 2005:** NREL collected production cost per watt as part of its program monitoring activity (Friedman et al., 2005). NREL's data collection protocols were reviewed and found to be consistent with data needs for this analysis.[2] These data were compared with pricing data from EIA and International Energy Agency (IEA) reports, which provided average module price by technology for 1992 to 2007, to estimate gross margins (EIA, 2008; IEA, 2009). The gross margin was estimated to be approximately 25 per cent in the 1990s.
- **2006 to 2008:** Data were provided by companies or estimated by financial reports.

PV systems are solid-state energy systems that have long lives. The minimum guaranteed lifetime for a module is 25 years, with the expectation that most modules will convert solar energy into electrical current for additional years. This concept is referred to as 'reliability'. The actual trend in guaranteed module lifetime was the baseline module reliability. Although module lifetimes may extend beyond the guaranteed period, guaranteed lifetime was the baseline, given that the producer incurs a

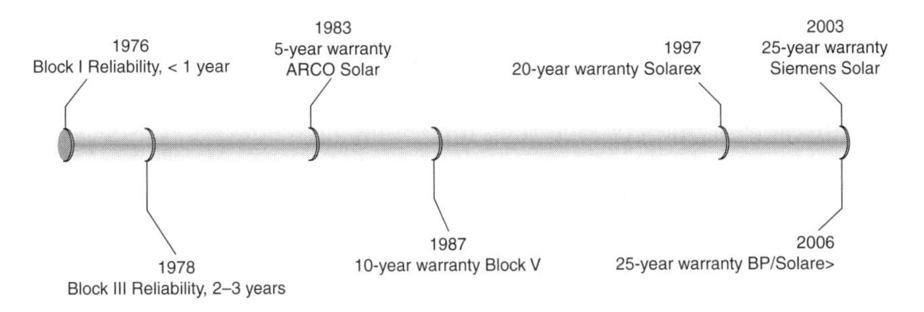

Sources: Christensen (1985); Green (2005).

Figure 5.2 Timeline of PV warranty introduction (guaranteed reliability)

financial consequence if that performance standard is not met. In 1982, Arco Solar (now SolarWorld USA) offered the first module warranty of five years.[3] At the close of FSA, ten-year warranties were offered, and modules were expected to last for 20 years. In the early 1990s, Solarex (now BP Solar) offered the first 20-year warranty. By the late 1990s, 25-year warranties had become standard (Figure 5.2).

Company-level production data (MW) for all DOE-funded companies from 1974 to 2008 were summed by year to generate an aggregate industry production quantity time series.[4] Data sources were as follows:

- **1974 to 1985**: FSA reports (Christensen, 1985).
- **1986 to 2004**: *PV News* (Maycock, 1986–2004; *PV News*, 2005–2009; Watts et al., 1984) (see also Margolis [2002]).
- **2005 to 2008**: EIA and IEA (EIA, 2008; IEA, 2009).

Although US companies are major players in the global market, the domestic market for photovoltaics is not as large as the market in other countries. The US ranked fourth in the world for PV installations in 2008, behind Germany, Spain, and Japan.[5] In the US, California is the leading state in PV capacity and accounts for more than 60 per cent of grid-connects PV installations (IEA, 2009).

The following were the data sources and estimation procedures for installations in the US:

- **1974 to 1984:** Little international trade in photovoltaics occurred, and all production for 1976 to 1981 was assumed to be installed domestically. Installations for 1981 through 1984 were estimated net of exports via simple regression based on EIA data for 1985 to 1992.
- **1985 to 1992:** Estimated installations were from EIA (2008).

- **1993 to 2008:** The annual change in installed photovoltaics in the US was derived from IEA market analyses (2009).

Although modules produced by foreign manufacturers are installed in the US, these modules must have met the cost and reliability specifications established by the DOE and US producers, and expected by US consumers. In light of the fact that foreign producers relied on FSA technology in the public domain and technology infrastructure supplied by DOE for their R&D and manufacturing processes, it was therefore reasonable for benefits calculations to include total US installations and not just installations of US-produced photovoltaics. This is referred to in economics as induced innovation.

The total installed cost of a system is considered along with the system's lifetime and power rating to yield the LCOE. The LCOE is usually presented as dollars per kilowatt-hour ($/kWh). The total energy produced is calculated by power multiplied by time and is reported by power producers as kilowatt-hours. The amount of energy in kilowatt-hours produced in a year by an electricity generator is the capacity in kilowatts multiplied by the number of hours in a year adjusted by a capacity factor to adjust for periods of nonoperation, or in the case of solar power, for when the sun is not shining at its peak. In the US, PV modules, on average, have a capacity factor of 18 per cent. Thus, one watt-peak (Wp) module can be expected to produce 1.58 kWh per year. Wp is a measure of power output under standard reporting conditions. Although the cost per kilowatt-hour can be estimated by taking into account the cost per watt, the lifespan of the module and the number of hours of available sunlight per day, the calculation result provides only a rough approximation.

Economic Models for Quantifying Economic Benefits

This study used two models for quantifying economic benefits. One quantified production cost savings alone for units that were not installed in the US. The other quantified benefits for installations in the US, which included both production cost savings and reliability benefits. Two models ensured that reliability gains accruing to non-US consumers were not included in the measures of economic return.

For modules installed in the US, the combined effect of simultaneous increases in reliability and reductions in cost yielded an amplified economic benefit greater than if one of these benefits occurred and the other did not. To monetize the benefits of improved reliability, a baseline annualized module cost was developed using the production cost per watt time series and the reliability curve, as shown in equation 5.1

$$AC = \frac{C}{\sum_{t=0}^{T} \frac{1}{(1+r)^t}}, \tag{5.1}$$

where

AC = annualized PV module cost (2008$)
C = PV module cost (2008$)
T = PV module lifetime (years)
r = discount rate.

This equation represents the annualized cost of a PV module that factors into the PV system purchase decision. Because this calculation is sensitive to the discount rate applied, to calculate measures of economic return, separate curves for each social discount rate of interest must be calculated.

A model to account for both a change in production cost and a change in expected lifetime was developed. Benefits were calculated using the following equation to compare the baseline (actual) module cost and reliability to the counterfactual module cost and reliability for the quantity installed in the US:

$$Q_{US} \times \left(\frac{C_c}{\sum_{t=0}^{T_c} \frac{1}{(1+r)^t}} - \frac{C_b}{\sum_{t=0}^{T_b} \frac{1}{(1+r)^t}} \right) \times \sum_{t=0}^{T_b} \frac{1}{(1+r)^t}, \tag{5.2}$$

where

Q_{US} = quantity of modules installed in the US in any given year (W)
C_c = counterfactual module production cost per watt ($/W)
C_b = actual module production cost per watt ($/W)
T_c = counterfactual module reliability (years)
T_b = actual module reliability (years)
r = discount rate.

Because reliability for modules installed outside the US is excluded, economic benefits are quantified simply as:

$$Q_{Non-US} \times (C_c - C_b), \tag{5.3}$$

where

$Q_{Non\text{-}US}$ = quantity of modules produced in any given year for the non-US market (W)

C_c = counterfactual module production cost per watt ($/W)
C_b = actual module production cost per watt ($/W).

ECONOMIC BENEFITS

Economic analysis results of the DOE's contributions to PV modules are presented in the following order:

1. Technology acceleration and counterfactual module reliability.
2. Technology acceleration and counterfactual production cost per watt.
3. Total economic benefits of higher-quality, lower-cost PV modules separated by (a) cost savings and reliability benefits for modules installed in the US, and (b) cost savings for modules destined for non-US markets.
4. Polysilicon production methods and the accelerated introduction of wire saw technologies

Technology Acceleration and Counterfactual PV Module Reliability

Between 1975 and 1985, FSA supported and integrated R&D efforts across every aspect of the terrestrial PV industry, from cell and module process improvements and engineering improvements to the incorporation of PV standards into the national electric code. During interviews experts often referred to the suite of technologies described in Section 3.1 and noted as evidence the extent to which those technologies are still embodied in commercial products.

Most experts estimated FSA's acceleration effect on cost reductions and reliability improvements to be between 10 and 15 years, with a whole-year average of 12 years.[6] A 12-year acceleration implies that the progress made over the 10 years of the FSA program would have instead taken 22 years. Figure 5.3 illustrates this effect's impact on guaranteed PV module reliability. Shifting milestones back 12 years places the introduction of the 5-year warranty in 1994 instead of 1982 and the introduction of the 20-year warranty in 2002 instead of 1990. Twenty-five-year warranties would not have been introduced within the period of analysis.

Counterfactual PV Module Production Cost Per Watt

The production-weighted average counterfactual production cost per watt curves depicted in Figure 5.4 were developed by aggregating company-specific responses to how their technology portfolios and manufacturing

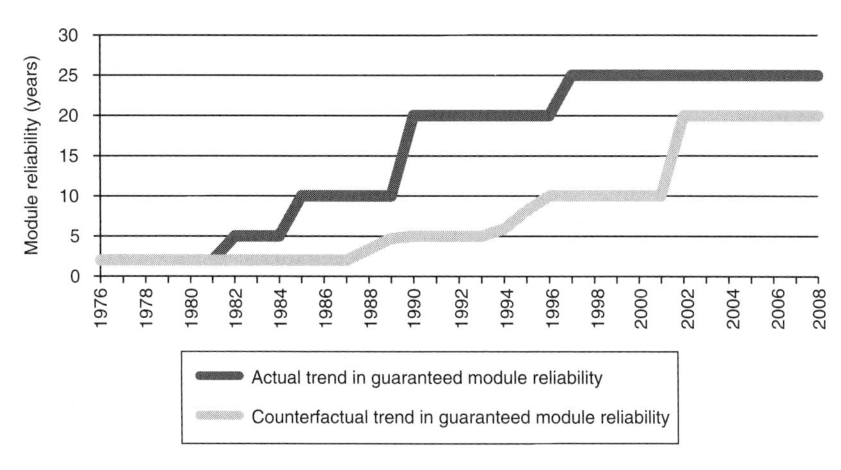

Sources: Authors' calculations.

Figure 5.3 Actual and counterfactual reliability curves

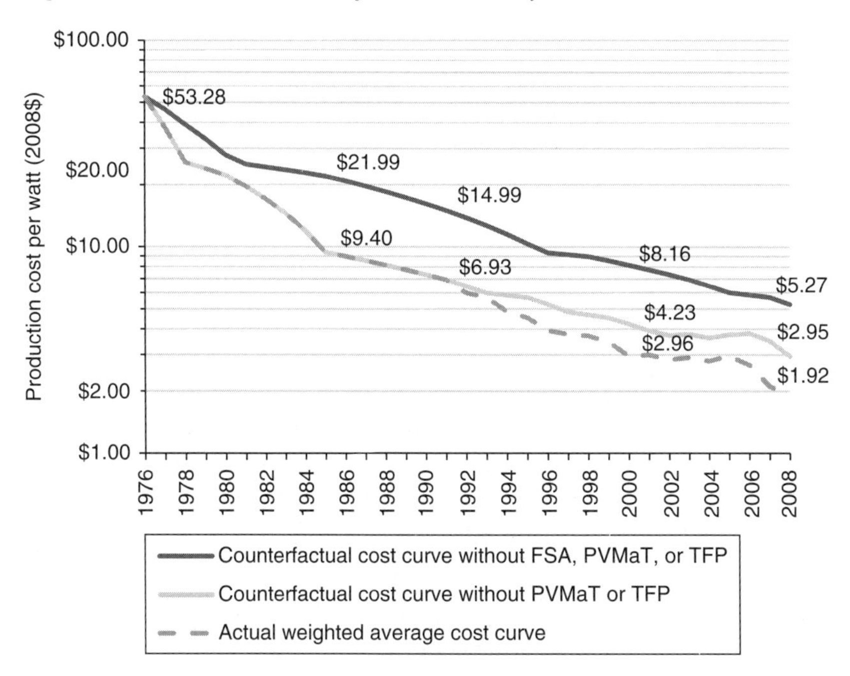

Sources: Authors' calculations.

Figure 5.4 Actual and counterfactual PV module production cost per watt curves (2008$)

operations would have developed in the absence of DOE cost sharing. Production cost-per-watt reductions (2008$) were greatly accelerated because of FSA, and technologies developed under PVMaT and TFP further hastened these reductions. Figure 5.4 presents three curves:

1. The dashed curve is the actual weighted average production cost per watt curve against which progress in the absence of DOE and its resources was measured. In 2008 US dollars, cost per watt was $9.40 in 1985, $6.93 in 1991, $2.96 in 2000, and $1.92 in 2008.
2. The dark-gray curve is the counterfactual, weighted average production cost per watt curve that presents the aggregate progress in the absence of DOE involvement, as determined by expert interviewees' assessment of DOE's impact. In the absence of DOE cost sharing, technical expertise, and technology infrastructure, industry progress would have proceeded at a slower pace. Note that in 1985, the last year of FSA, the cost per watt would have been $21.99, rather than $9.40. In 2008, it would have been $5.27, rather than $1.92—a difference of $3.35 per watt.
3. The light–gray curve beginning in 1991 illustrates the effect of PVMaT and TFP. If PVMaT and TFP had not followed FSA, then from 1991 onward the cost per watt would have diverged from the green path to the red path. Costs would have been as much as 66 per cent higher, the rate of progress would have been lower, and the weighted average cost would have been $2.95 in 2008 rather than $1.92.

In 2008, the difference between the actual and counterfactual cost was $3.35 per watt, of which $2.32 was associated with the acceleration effect from FSA and $1.03 was associated with PVMaT and TFP technology.

This gap in, and the differences between, DOE initiatives, combined with the start-up of new PV companies, translated into experts suggesting DOE's influence be segmented by initiative: FSA and PVMaT/TFP combined. The last block purchase under FSA was in 1984, and the project ended in 1985. Recall that PVMaT and TFP, which ran largely concurrently, did not ramp up significantly until 1992 to 1994. This segmentation also enables benefits to be separated between FSA's foundational research into all aspects of photovoltaics and those from targeted R&D into manufacturing systems (PVMaT) and thin films (TFP and PVMaT).

PVMaT and TFP were combined for two reasons. First, only a few US PV companies participated in these programs and had PV module production and cumulative installations for which economic benefits were quantified. Combining the programs precluded disclosure of individually identifiable results.

Second, many companies received funding under both programs, and although they were able to assign technical impacts between programs, the interplay between economic impacts from the two programs and the rapid scale-up of the thin-film sector was such that gains from PVMaT and TFP individually could not be distinguished meaningfully.

Table 5.3 presents the data from Figure 5.4 (in tabular format as well as with the percentage increase of counterfactual over actual production cost per watt) to document interviewees' aggregate responses. Note that in the case of PVMaT and TFP alone, cost per watt would have been in the range of approximately 20 per cent to 60 per cent over the actual cost between 1994 and 2005. The difference is greatest for 2007 and 2008 as thin-film technologies entered the market in large numbers because of these technologies' production cost advantages over c-Si, despite the lower average energy conversion efficiency.

TOTAL ECONOMIC BENEFITS FROM HIGHER-QUALITY, LOWER-COST MODULES

Total economic benefits attributable to the DOE from technology acceleration and development, as reflected by reductions in production cost per watt and gains in reliability presented in the preceding sections, are estimated to have been $18,093 million between 1976 and 2008, in real terms (2008$).[7]

For PV modules installed in the US between 1976 and 2008, the DOE-supported technology offset what would otherwise be higher production costs per watt and lower guaranteed module lifetime. The simultaneous accrual of production cost reductions and reliability gains generated benefits of $11,320 million (Table 5.4), a figure that would have been lower had only cost savings or only reliability gains been achieved.

Note in particular the period between 1985 and 1987. Although annual installations only ranged in the upper 5 MW, benefits were estimated to be over $500 million (2008$) in each year. The influence is attributable to accelerated cost reductions and reliability gains achieved under FSA. Whereas consumers were installing modules in 1984 with a ten-year guaranteed lifetime, in the absence of FSA those modules would only have a two-year expected lifetime. A module installed 10 years later in 1994 had a 20-year expected lifetime. This is an empirical result of the acceleration effect. Study participants indicated a 12-year acceleration of the industry along all major measures of progress under FSA. The 1986 cost per watt was delayed until 1998, and without PVMaT or TFP, the rate of progress for 1998 to 2008 would have corresponded roughly to actual progress for the late 1980s and early 1990s.

Table 5.3 Actual and counterfactual weighted-average production cost per watt (2008$)

Year	Actual production cost per watt ($/W)	Counterfactual cost without FSA and PVMaT/TFP ($/W)	% Increase over baseline	Counterfactual cost without PVMaT/TFP ($/W)	% Increase over baseline
1974	114.44	114.44	–	–	–
1975	83.86	83.86	–	–	–
1976	53.28	53.28	–	–	–
1977	37.60	46.15	23	–	–
1978	25.64	39.03	52	–	–
1979	23.93	33.25	39	–	–
1980	22.22	27.81	25	–	–
1981	19.65	25.17	28	–	–
1982	17.09	24.39	43	–	–
1983	14.53	23.62	63	–	–
1984	11.96	22.84	91	–	–
1985	9.40	21.99	134	–	–
1986	8.99	20.82	132	–	–
1987	8.58	19.65	129	–	–
1988	8.16	18.49	126	–	–
1989	7.75	17.32	123	–	–
1990	7.34	16.16	120	7.34	–
1991	6.93	14.99	116	6.93	–
1992	6.00	13.83	131	6.46	8
1993	5.69	12.66	122	6.00	5
1994	4.84	11.50	138	5.85	21
1995	4.53	10.33	128	5.69	26
1996	3.93	9.36	138	5.27	34
1997	3.77	9.18	143	4.84	28
1998	3.71	8.99	142	4.68	26
1999	3.45	8.58	148	4.53	31
2000	2.96	8.16	176	4.23	43
2001	3.00	7.75	159	3.93	31
2002	2.85	7.34	158	3.73	31
2003	2.91	6.93	138	3.77	30
2004	2.80	6.46	131	3.63	30
2005	2.96	6.00	102	3.76	27
2006	2.67	5.85	119	3.81	43
2007	2.11	5.69	170	3.50	66
2008	1.92	5.27	174	2.95	54

Sources: Authors' calculations.

Table 5.4 Economic benefits from PV modules installed in the United States (2008$)

Year	Annual US PV installed (MW)	Actual		Counterfactual		Incremental change		Economic benefit ($ million)
		Cost ($/W)	Reliability (Years)	Cost ($/W)	Reliability (Years)	Cost ($/W)	Reliability (Years)	
1976	0.8	53.28	2	53.28	2	–	–	–
1977	1.2	37.60	2	46.15	2	8.55	–	10.5
1978	1.6	25.64	2	39.03	2	13.39	–	22.1
1979	2.1	23.93	2	33.25	2	9.32	–	19.3
1980	2.5	22.22	2	27.81	2	5.59	–	14.0
1981	4.5	19.65	2	25.17	2	5.52	–	24.6
1982	5.0	17.09	5	24.39	2	7.30	3	221.4
1983	5.2	14.53	5	23.62	2	9.09	3	231.7
1984	5.4	11.96	5	22.84	2	10.88	3	242.1
1985	5.5	9.40	10	21.99	2	12.59	8	555.3
1986	5.7	8.99	10	20.82	2	11.83	8	540.5
1987	5.8	8.58	10	19.65	2	11.07	8	524.1
1988	6.0	8.16	10	18.49	2	10.33	8	280.9
1989	6.2	7.75	10	17.32	2	9.57	8	178.0
1990	6.3	7.34	20	16.16	2	8.82	18	362.2
1991	6.5	6.93	20	14.99	2	8.06	18	343.8
1992	6.6	6.00	20	13.83	2	7.83	18	327.5
1993	6.8	5.69	20	12.66	2	6.97	18	305.7
1994	7.5	4.84	20	11.50	5	6.66	15	255.6
1995	9.0	4.53	20	10.33	5	5.80	15	186.6
1996	9.7	3.93	20	9.36	5	5.43	15	143.5
1997	11.7	3.77	25	9.18	10	5.41	15	224.2
1998	11.9	3.71	25	8.99	10	5.28	15	223.3
1999	17.2	3.45	25	8.58	10	5.13	15	309.4
2000	21.5	2.96	25	8.16	10	5.20	15	375.3
2001	29.0	3.00	25	7.75	10	4.75	15	475.1
2002	44.4	2.85	25	7.34	20	4.49	5	281.0
2003	63.0	2.91	25	6.93	20	4.02	5	362.2
2004	100.8	2.80	25	6.46	20	3.66	5	532.2
2005	103.0	2.96	25	6.00	20	3.04	5	466.6
2006	145.0	2.67	25	5.85	20	3.18	5	672.0
2007	206.5	2.11	25	5.69	20	3.58	5	1 034.8
2008	338.0	1.92	25	5.27	20	3.35	5	1 574.1
Total								11 319.5

Source: Authors' calculations.

Larger annual economic benefits began to accrue in 2003 as the volume of photovoltaics installed in the US every year increased from 63 MW in 2003 to 338 MW in 2008. Legacy impacts attributable to foundational technologies developed under FSA were combined with the introduction of thin-film technologies, which offer lower material costs to producers at the expense of energy conversion efficiency, and operational efficiency achieved through the introduction of advanced manufacturing technologies. The difference in guaranteed reliability for these years was five years, but increasingly large volumes of photovoltaics installed multiplied by accelerated introduction of cost per watt indicate that DOE's role in technology development increasingly delivered value to consumers. That value was estimated at $1,574 million in 2008 alone.

The rightmost columns in Table 5.4 present the discounted time series of economic benefits. Reliability analyses are sensitive to discount rates, as the economic model indicates. This is because when consumers purchase a module today, they are looking at the LCOE they expect to lock in over the lifetime of their module. Because modules are a sunk cost for consumers, the reliability benefit is part of the investment decision and can, therefore, be treated as a one-time gain that distributes noncash benefits over time. To calculate measures of return accurately, the same discount rates for LCOE assessments must be used as for the measures of return.

PV companies receiving DOE cost shares produced a large volume of modules destined for non-US markets. Accordingly, the analysis only valued benefits from production cost savings accruing to producers, excluding the reliability benefits accruing to non-US consumers. Multiplying the annual cost difference by the subset of production volume yields $6,773 million over the period from 1976 to 2008 (Table 5.5).

ECONOMIC BENEFITS OF UCC POLYSILICON PRODUCTION METHOD

The FSA project's silicon material initiative's goal was to generate polycrystalline silicon feedstock at a reduced cost to the traditional trichlorosilane Siemens process. FSA contractors explored several different processes; however, only the process developed by UCC became deployed for commercial production. The UCC process uses silane gas as opposed to trichlorosilane as a feedstock to deposit polycrystalline silicon using the Siemens process. Advantages of the UCC process include 'a lower deposition-reaction temperature, a higher conversion efficiency, and lower environmental and corrosion problems' (Lutwack, 1986).

Detailed engineering analyses at the time indicated that the UCC process

Table 5.5 Economic benefits of PV modules destined to non-US markets (2008$)

Year	Economic benefit of US installations ($ million)	Production (MW)	Incremental cost savings benefits ($/W)	Economic benefits ($ million)	Total module technology economic benefits ($ million)
1976	–	0.00	0.00	–	–
1977	10.5	0.00	8.55	–	10.5
1978	22.1	0.00	13.39	–	22.1
1979	19.3	0.00	9.32	–	19.3
1980	14.0	0.00	5.59	–	14.0
1981	24.6	0.00	5.52	–	24.6
1982	221.4	0.00	7.30	–	221.4
1983	231.7	0.43	9.09	3.9	235.5
1984	242.1	0.90	10.88	9.8	251.9
1985	555.3	2.28	12.59	28.6	583.9
1986	540.5	1.57	11.83	18.5	559.1
1987	524.1	3.01	11.08	33.3	557.4
1988	280.9	5.55	10.33	57.3	338.2
1989	178.0	8.24	9.57	78.9	256.9
1990	362.2	8.83	8.82	77.9	440.0
1991	343.8	11.00	8.07	88.7	432.5
1992	327.5	11.96	7.83	93.7	421.2
1993	305.7	15.64	6.97	109.0	414.7
1994	255.6	18.76	6.66	124.9	380.5
1995	186.6	25.96	5.80	150.7	337.2
1996	143.5	30.11	5.44	163.6	307.2
1997	224.2	39.40	5.40	212.8	437.1
1998	223.3	42.00	5.28	221.7	445.0
1999	309.4	43.60	5.12	223.3	532.7
2000	375.3	53.50	5.21	278.6	653.9
2001	475.1	71.50	4.75	340.0	815.1
2002	281.0	83.20	4.49	373.9	654.8
2003	362.2	39.62	4.02	159.2	521.4
2004	532.2	37.90	3.66	138.9	671.1
2005	466.6	75.10	3.04	227.7	694.3
2006	672.0	122.80	3.18	389.7	1061.7
2007	1034.8	245.70	3.58	881.4	1916.2
2008	1574.1	684.60	3.35	2287.0	3861.1
Total	11319.5			6773.0	18092.5

Source: Authors' calculations.

Table 5.6 Economic benefits from UCC polycrystalline silicon production process (2008$)

Year	Capacity (million tons)	Production (million tons)	Cost savings per kilogram ($/kg)	Economic benefits ($ million)
1985	1000	800	8.53	7.2
1986	1000	800	8.53	7.2
1987	1400	1120	8.53	10.1
1988	1400	1120	8.53	10.1
1989	1400	1120	8.53	10.1
1990	1400	1120	8.53	10.1
1991	1400	1120	8.53	10.1
1992	1400	1120	8.53	10.1
1993	1400	1120	8.53	10.1
1994	1400	1120	8.53	10.1
1995	1400	1120	8.53	10.1
1996	2100	1680	8.53	15.2
1997	2100	1680	8.53	15.2
1998	5900	4720	8.53	42.8
1999	5900	4720	8.53	42.8
2000	5900	4720	8.53	42.8
2001	5900	4720	8.53	42.8
2002	5900	4720	8.53	42.8
2003	5900	4720	8.53	42.8
2004	5900	4720	8.53	42.8
2005	5900	5300	8.53	45.2
2006	5900	5555	8.53	47.4
2007	6900	5780	8.53	49.3
2008	6900	6171	8.53	52.6
Total				630.1

Sources: Flynn and Bradford (2006); authors' calculations.

was less expensive by $8.53/kg (2008$) (Yaws et al., 1986). For the purposes of this analysis, it was assumed that, although many of the production costs may have fluctuated over time, this cost difference has remained constant.[8]

Table 5.6 presents data on plant production capacity based on the UCC process, estimated annual output, and the cost savings accruing from using this novel silicon refining process from 1985 to 2008. The first commercial facility implementing the UCC process went into production in 1985 in Moses Lake, Washington, with a capacity of 1,000 MT. This capacity was expanded to 1,400 MT in 1987 and again to 2,100 MT in 1996 (Flynn and Bradford, 2006). In 1990, Kanetsu acquired the production facility from UCC and renamed it Advanced Silicon Materials Inc. (ASiMI). ASiMI constructed an additional facility with a capacity of 3,800 MT in Butte, Montana, that came into production in 1998. In 2002, the Renewable Energy Corporation (REC) developed a joint venture with ASiMI at the Moses Lake facility to form Solar Grade Silicon (SGS). In 2005, REC acquired the Butte, Montana, facility from ASiMI and began a debottlenecking project to increase polysilicon capacity by 1,000 MT. Maximum capacity for 2007 and 2008 was, therefore, 6,900 MT.

Capacity utilization rates between 2005 and 2008 averaged 89 per cent. There was an oversupply of silicon in 2007, and the facilities operated at only 84 per cent capacity that year. An average utilization rate of 90 per cent was used to estimate production for 1985 to 2004 because actual production information was unavailable. Through the end of 2008, approximately 70,900 MT of polysilicon had been produced using the UCC process. This amounts to a total benefit of $630 million (2008$).

ECONOMIC BENEFITS OF ACCELERATED INTRODUCTION OF WIRE SAW TECHNOLOGY TO THE SEMICONDUCTOR INDUSTRY

The wire saw is a less costly technology for slicing silicon ingots relative to the alternative defender technology, internal diameter saws. Wire saws are capable of cutting larger silicon ingots into smaller wafers with less kerf loss (wasted silicon) than internal diameter saws. In addition, they are capable of cutting an entire ingot into wafers at once. The development and adoption of wire saws for silicon slicing was driven primarily by the requirement for low-cost silicon wafers by the PV industry. Costs associated with slicing silicon are small relative to the value added by wafer manufacturers in the semiconductor industry. However, the need for inexpensive silicon wafers for photovoltaics compelled the industry and DOE to explore wire saws.

Wire saws were assessed in the 1970s and 1980s as part of FSA. Although beneficial for the reasons described above, these wire saws had high variable and maintenance costs and, thus, were not cost-effective. In the early 1990s, PVMaT enabled both Solarex and Siemens to assess contemporary wire saw machinery, and both companies successfully adopted wire saws in 1993. Solarex replaced all 24 of their internal diameter saws with a single wire saw and purchased several more wire saws. Solarex reported that the wire saws provided a savings of $0.13 per wafer.

Expert interviewees have indicated that without DOE funding, wire saws

Table 5.7 Economic benefits from accelerated adoption of wire saws in the semiconductor industry (2008$)

Year	200 mm wafers (million square inches)	300 mm wafers (million square inches)	Cost savings ($ million)	Cost savings under delayed introduction ($ million)	Economic benefits of accelerated wire saw introduction ($ million)
1994	7824247	–	<0.0	–	<0.0
1995	17640493	–	0.4	–	0.4
1996	25875462	–	2.3	–	2.3
1997	33658636	9127	3.8	0.1	3.7
1998	33453275	45636	4.2	0.9	3.4
1999	41462347	456357	5.3	3.7	1.6
2000	57952820	857952	7.5	6.7	0.7
2001	47171377	1168275	6.1	6.1	–
2002	50929480	3194501	6.9	6.9	–
2003	59000160	4034199	8.0	8.0	–
2004	68816406	7246955	9.6	9.6	–
2005	66085107	12221250	9.9	9.9	–
2006	72800406	19167009	11.7	11.7	–
2007	69001231	28339791	12.3	12.3	–
2008	54934016	32373990	11.1	11.1	–
Total			99.3	87.1	12.2

Sources: SEMI (2009); authors' calculations.

would have eventually been adopted by the PV industry and subsequently the semiconductor industry. DOE funding accelerated the adoption of wire saws to the semiconductor industry by an estimated three years.[9] We assume that wire saw adoption in 200 mm wafer production followed an S-shaped adoption pattern in the semiconductor industry from 1994 to 1998. Based on an estimate of $0.13 saved per wafer, total economic benefits of wire saws were estimated to be $99.3 million (Table 5.7). Benefits attributable to the DOE from accelerated introduction of this technology to the industry were estimated to be $12.2 million.

BENEFIT-COST ANALYSIS OF PHOTOVOLTAIC ENERGY SYSTEMS CLUSTER

This section presents the summary total of economic benefits, measures of economic performance, and sensitivity analysis for the technology cluster Photovoltaic Energy Systems.

Measures of Economic Return for the Technology Cluster

Total quantified economic benefits were compared with the total public investment in Photovoltaic Energy Systems to provide lower-bound measures of economic return for the entire cluster. Between 1975 and 2008 Congress had appropriated $7,438 million for solar energy, including $3,710 million for Photovoltaic Energy Systems (2008$).

Economic benefits were estimated to be $18,735 million, of which:

- $18,092 million was accrued from higher-quality, lower-cost PV modules
- $630 million was accrued from the UCC polycrystalline silicon refinement process
- $12 million was accrued from the accelerated adoption of wire saw technology by the semiconductor industry.

Thus, net of investment costs of $3,710 million, net economic benefits were $15,025 million in real terms (Table 5.8). The IRR was 17 per cent. Also, applying a 7 per cent discount rate yields an NPV of $1,459 million and a benefit-cost ratio (BCR) of 1.83.[10] Applying a 3 per cent discount rate yields an NPV of $5,725 million and a BCR of 3.24.

Measures of Return for FSA and for PVMaT/TFP

To review long-term influences, this study also reorganized economic benefit results by initiative (Table 5.9):

- FSA ran from 1975 to 1985, cost the DOE $535 million, and continues to generate economic benefits, which through 2008 amounted to $15,673 million. Applying the 7 per cent social discount rate provides a BCR of 7.12 and an NPV of $2,435 million. The IRR was 37 per cent.
- PVMaT and TFP ran from 1988 to 2008, cost the DOE $495 million, and continues to generate economic benefits, which through 2008 amounted to $3,062 million. Applying the 7 per cent social discount rate provides a BCR of 3.35 and an NPV of $637 million. The IRR was 24 per cent.

That the IRRs of FSA and PVMaT/TFP were individually greater than the cluster IRR of 17 per cent results from including cluster costs for which no benefits were calculated in the time series of cash flows.

It is also important to note that benefits for FSA accrued over the

Table 5.8 Lower-bound net economic benefits from DOE investment in photovoltaic energy systems (2008$)

Year	Module technology benefits ($ million)	UCC polysilicon production process ($ million)	Accelerated adoption of wire saw technology ($ million)	Total benefits ($ million)	Total costs, photovoltaic energy systems ($ million)	Net benefits ($ million)
1975	–	–	–	–	(1.9)	(1.9)
1976	–	–	–	–	(65.9)	(65.9)
1977	10.5	–	–	10.5	(170.7)	(160.2)
1978	22.1	–	–	22.1	(204.6)	(182.5)
1979	19.3	–	–	19.3	(294.5)	(275.2)
1980	14.0	–	–	14.0	(340.9)	(326.9)
1981	24.6	–	–	24.6	(314.9)	(290.3)
1982	221.4	–	–	221.4	(144.9)	76.6
1983	235.5	–	–	235.5	(109.1)	126.5
1984	251.9	–	–	251.9	(91.1)	160.8
1985	583.9	7.2	–	591.2	(96.3)	494.9
1986	559.1	7.2	–	566.3	(69.5)	496.9
1987	557.4	10.1	–	567.5	(67.4)	500.1
1988	338.2	10.1	–	348.4	(56.2)	292.2
1989	256.9	10.1	–	267.1	(54.8)	212.2
1990	440.0	10.1	–	450.2	(51.6)	398.6
1991	432.5	10.1	–	442.7	(66.8)	375.8
1992	421.2	10.1	–	431.3	(85.0)	346.3
1993	414.7	10.1	–	424.8	(90.0)	334.8
1994	380.5	10.1	<0.0	390.7	(101.7)	289.0
1995	337.2	10.1	0.4	347.8	(111.5)	236.3
1996	307.2	15.2	2.3	324.7	(80.0)	244.7
1997	437.1	15.2	3.7	456.0	(76.0)	380.0
1998	445.0	42.8	3.4	491.2	(82.1)	409.1
1999	532.7	42.8	1.6	577.0	(88.2)	488.8
2000	653.9	42.8	0.7	697.4	(79.0)	618.4
2001	815.1	42.8	–	857.8	(88.9)	768.9
2002	654.8	42.8	–	697.6	(77.1)	620.5
2003	521.4	42.8	–	564.2	(84.4)	479.7
2004	671.1	42.8	–	713.9	(81.3)	632.5
2005	694.3	45.2	–	739.5	(71.4)	668.0
2006	1 061.7	47.4	–	1 109.1	(34.0)	1 075.0
2007	1 916.2	49.3	–	1 965.5	(141.3)	1 824.1
2008	3 861.1	52.6	–	3 913.7	(136.7)	3 776.9
Total	18 092.5	630.1	12.2	18 734.8	(3 709.9)	15 024.9

Source: Authors' calculations.

Table 5.9 Lower-bound measures of economic return for photovoltaic energy systems

Measure	Photovoltaic energy systems cluster	FSA (1975–1985)	PVMaT (1991–2008) TFP (1988–2008)
Period of net benefits accrual	*1975–2008*	*1975–2008*	*1988–2008*
Total benefits (million 2008$)	$18734.8	$15673.3	$3061.5
Total costs (million 2008$)	$3709.9	$535.0	$495.0
Net benefits (million 2008$)	$15024.9	$15138.3	$2556.6
Internal rate of return	17%	37%	24%
NPV at 7% (million 2008$)	$1458.9	$2435.1	$636.9
Benefit-to-cost ratio at 7%	1.83	7.12	3.35
NPV at 3% (million 2008$)	$5724.7	$6592.8	$1409.9
Benefit-to-cost ratio at 3%	3.24	15.07	4.76

Source: Authors' calculations.

entire 33-year period of analysis. Results for PVMaT and TFP reflect more recent investments, and economic returns from the DOE's investment in thin-film PV in particular, are only now beginning to accrue. Chapter 4 highlights that large-volume production of thin-film PV did not begin until 2003, but investment was sustained by the DOE, beginning in 1988. This constituted a nearly 15-year incubation period. Thus, it is expected that the annual public return on investment in PVMaT and TFP will exceed the 24 per cent calculated for the 20-year period from 1988 to 2008.

Sensitivity Analysis on Measures of Economic Return for the Photovoltaic Energy Systems Cluster

Cash-flow analyses are sensitive to the timing of cash flows, and this study spanned 33 years of DOE investment and identified a significant technology acceleration effect. The earlier a cash flow accrues in a series, the greater its influence on the measure. Thus, a sensitivity analysis was performed on the measures of economic return by calculating how calculated values would change under alternative acceleration periods. FSA's

Table 5.10 Sensitivity analysis of FSA acceleration effect on economic performance measures

Measure	Results (12-year acceleration)	Under 10-year FSA Acceleration effect	Under 15-year FSA Acceleration effect
Total benefits (million 2008$)	$18 734.8	$14 389.8	$25 875.7
Total costs (million 2008$)	$3 707.9	$3 707.9	$3 707.9
Net benefits (million 2008$)	$15 026.8	$10 681.8	$22 167.7
Internal rate of return	17%	14%	20%
NPV at 7% (million 2008$; base year = 1975)	$1 458.9	$858.8	$2 394.6
Benefit-to-cost ratio at 7%	1.83	1.49	2.37
NPV at 3% (million 2008$; base year = 1975)	$5 724.7	$3 987.2	$8 531.5
Benefit-to-cost ratio at 3%	3.24	2.56	4.35

Source: Authors' calculations.

technology acceleration effect had the most significant effect on the industry weighted-average counterfactual production cost per watt (also see Figure 5.4).

Recall from the 'Model for estimating economic benefits' section that the average acceleration effect incorporated into the counterfactual cost per watt curve was 12 years. The distribution of experts' quantitative estimates was between 10 and 15 years. Therefore, the sensitive analysis calculated two alternative counterfactual cost curves: one for a 10-year acceleration effect and another for a 15-year effect (Table 5.10).

If the acceleration effect from FSA were 10 years rather than 12 years:

- Total benefits would have been $14,390 million and net benefits would have been $10,682 million.
- Applying a discount rate of 7 per cent, the NPV would have been $859 million, which is 41 per cent less than $1,459 million. Similarly, the BCR would have been 1.49 instead of 1.83.
- The IRR would have been 14 per cent.

If the acceleration effect from FSA were 15 years rather than 12 years:

- Total benefits would have been $25,876 million and net benefits would have been $22,168 million.

- Applying a discount rate of 7 per cent, the NPV becomes $2,395 million, which is 64 per cent more than $1,459 million. Similarly, the BCR would have been 2.37 instead of 1.83.
- The IRR would have been 20 per cent.

ENVIRONMENTAL AND ENVIRONMENTAL HEALTH BENEFITS

Electricity from PV systems, unlike fossil fuels and other sources of electricity, does not present environmental costs during energy generation. Yet the module production process is energy intensive. Before FSA, modules often failed within a year. By the end of the project, companies were offering ten-year warranties. Today, modules generally have a guaranteed lifetime of 25 years. This improved reliability reduces the number of modules that must be disposed of and replaced.

Failed PV modules are often disposed of in landfills because most can be safely thrown away. Recycling may not be cost-effective for companies because modules contain a relatively small amount of semiconductor and are widely dispersed among customers (EERE, 2009b). Yet some companies agree to take back modules at the end of their lifetime, for recycling. Of course, with longer lifetimes, the company may no longer be in existence by the time the module fails. Longer lifetimes ensure that fewer modules will end up as landfill, even if they are not recycled. CIS and CdTe thin-film modules contain toxic chemicals, and longer lifetimes help reduce the risk that these chemicals will leak out through improper disposal.

Environmental emissions benefits were estimated by comparing the reduction in air-pollutant emissions from using PV systems with the reductions using the next-best technology alternative. Electricity generation is a major source of GHG and other air-pollutant emissions. Solar energy, which serves as a substitute for GHG-producing energy sources such as natural gas, coal, and petroleum, does not release GHGs during energy production. The GHGs produced during electricity generation from fossil fuels are CO_2, methane (CH_4), and nitrous oxide (N_2O). Other GHGs, such as hydrofluorocarbons, perfluorocarbons and sulfur hexafluoride, are not directly associated with fossil fuel combustion and are, therefore, not included in this analysis.

PV installations were segmented by (1) grid-connected centralized, (2) grid-connected decentralized, and (3) off-grid applications. Each segment was then compared to the most likely fuel choice for the application, excluding any solar technologies. The percentage change in emissions

Table 5.11 Emissions underlying environmental health effects (avoided emissions [lbs/kWh])

	CO_2	CH_4	N_2O	PM	SO_2	NO_x	NH_3	VOCs
On-grid distributed	1.226208	0.000041	0.000010	0.000094	0.001022	0.000495	0.000004	0.000016
On-grid centralized	1.160000	0.000075	0.000101	0.000016	0.000006	0.000162	0.000004	0.000013
Off-grid	2.150000	0.000021	0.000276	0.000522	0.000332	0.009694	0.000004	0.000526

Sources: EIA (2009a). See also http://www.epa.gov/ttn/chief/ap42/ch03/final/c03s04.pdf.

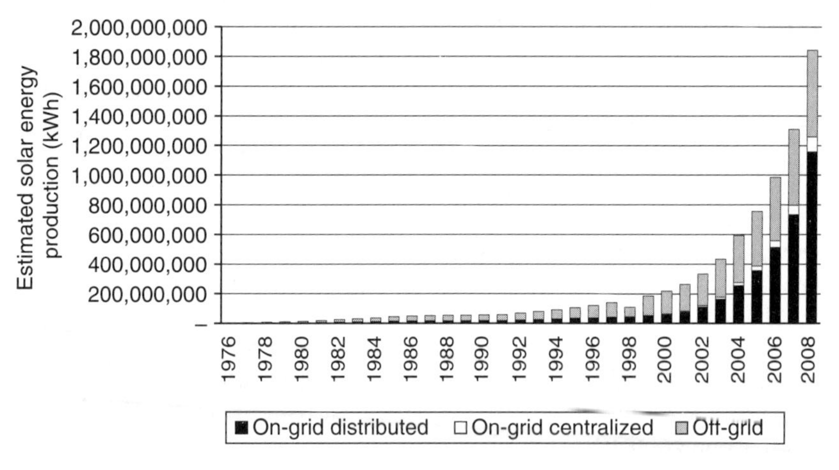

Source: EIA (2009b).

Figure 5.5 Solar energy production (kWh)

factors for electricity production, such as particulate matter (PM), nitrogen oxide (NO_x), sulfur dioxide (SO_2) and volatile organic compounds (VOCs), drove the model results (see Table 5.11 and Figure 5.5).

1. Grid-connected centralized applications are energy centers that are not associated with a particular customer and are primarily utility scale. The versatility and short ramp-up time of natural gas electricity generation units compare closely with PV systems, and these units are often used for peak hours, particularly during warmer months when air conditioning use increases electricity demand.

Table 5.12 Solar energy generation and average fossil fuel mix by state, 2008

State	Solar (kWh)	Coal	Natural gas	Petroleum
California	688 718 789	2%	96%	2%
New Jersey	91 516 296	35%	64%	2%
Colorado	46 540 338	70%	29%	0%
Nevada	44 584 862	24%	76%	0%
Arizona	32 982 369	52%	48%	0%
New York	28 549 955	28%	61%	11%
Hawaii	17 599 288	15%	0%	85%
Connecticut	11 472 128	25%	66%	9%
Oregon	10 038 112	23%	77%	0%
North Carolina	6 127 159	94%	5%	1%
Other	62 835 975	68%	30%	2%
Weighed fuel use		16%	80%	3%

Source: EIA (2009b).

Geothermal and wind power were not considered close substitutes to solar power, despite being renewable sources, because geothermal is considered base load, and the location profile favoring wind power may not align with that for solar. Thus, barring advances in storage technologies, PV electricity generation is limited to daytime conditions, and the nearest substitute for a PV system is a natural gas peaking unit.

The kilowatt-hours avoided for natural gas and internal combustion engines were multiplied by the emission factors for CO_2, CH_4 and N_2O available from the EIA (EIA, 2009a). For on-grid distributed applications, GHG emissions avoided were found by using the emissions factors specific to each region from EPA's eGRID and weighted by the kilowatt-hours of solar energy in each region (EPA, 2009a).

2. On-grid distributed applications refer to PV systems that are connected to the grid and used to provide power to a particular customer, such as a residential roof-mounted system. The power displaced by distributed PV systems depends on the fuel generation portfolio for each state. Current PV installations by state and the proportion of each fossil fuel type providing power in each state were reviewed (see Table 5.12). A weighted average kWh reduction was calculated for each fuel source: 80 per cent of emissions reductions came from the avoided use of natural gas, 16 per cent from coal, and 3 per cent from petroleum.

Table 5.13 Estimated avoided GHG emissions, 1976–2008

	Total avoided emissions			Approximate attribution to DOE		
	CO_2 (tons)	CH_4 (tons)	N_2O (tons)	CO_2 (tons)	CH_4 (tons)	N_2O (tons)
On-grid centralized	202 694	7	3	32 152	1	<1
On-grid distributed	2 346 139	83	33	372 154	13	5
Off-grid	4 266 270	42	548	658 167	6	84
Total	6 815 103	132	583	1 062 473	21	90

Source: COBRA estimates.

3. Off-grid applications were compared with diesel-fired internal combustion engines. Many remote off-grid PV modules, such as those on street signs or remote sheds, can take the place of a diesel generator. Because diesel generators produce more emissions per kilowatt-hour than natural gas, off-grid solar produced much larger benefits than on-grid solar, despite producing fewer kilowatt-hours in 2008.

Table 5.13 shows estimated total GHG emissions avoided.[11] Although fossil fuel combustion releases much smaller amounts of CH_4 and N_2O than CO_2, these GHGs are not trivial because they are approximately 21 times and 310 times, respectively, more effective at trapping heat in the atmosphere than CO_2 (EPA, 2009b). Avoided GHG emissions attributable to DOE were approximated through an acceleration analysis of efficiency improvement. The percentage of benefits attributable to DOE was approximated using the ratio of the baseline to the counterfactual in each year.

The EPA's Greenhouse Gas Equivalency Calculator converts GHG emissions to everyday terms (EPA, 2009c). Approximate equivalencies for total emissions avoided by photovoltaics in 2008 alone are the following:

- GHG emissions from 247,139 passenger vehicles.
- One year of CO_2 emissions from electricity use in 167,862 homes.
- Carbon sequestered annually from 275,595 acres of pine or fir forest.

The use of photovoltaics also avoids other harmful non-GHG emissions released during electricity production from coal, natural gas, oil, and other

Table 5.14 Estimated other emissions, 1976–2008

	Total avoided emissions				Approximate attribution to DOE			
	PM (tons)	SO_2 (tons)	NH_3 (tons)	VOCs (tons)	PM (tons)	SO_2 (tons)	NH_3 (tons)	VOCs (tons)
On-grid centralized	3	3	1	2	<1	<1	<1	<1
On-grid distributed	181	1964	8	31	32	352	1	6
Off-grid	1049	667	8	1057	174	111	1	175
Total	1232	2634	16	1090	207	463	3	181

Source: Authors' calculations.

combustibles. Emissions such as PM, ammonium (NH_3), and VOCs can have a negative impact on public health and the environment. Table 5.14 displays estimated emissions avoided and the amount of these attributable to the DOE.

ENVIRONMENTAL HEALTH BENEFITS

EPA's Co-Benefits Risk Assessment (COBRA) model was used to calculate the health benefits of reductions in air pollutants resulting from using PV systems. The COBRA model produces the incidence and cost of health effects. Incidence is defined as the change in number of health incidents relative to natural gas combustion.

According to COBRA, for 2008 alone, avoided adverse health incidents were estimated to be as follows:

- On-grid centralized systems: $90,500 for 100.1 million kWh (Table 5.15).
- Grid-connected distributed systems: $11.8 million for 1,158.9 million kWh (Table 5.15).
- Off-grid systems: $28.7 million in 2008 for 583.4 million kWh (Table 5.16).

Thus, for 2008 alone, the total environmental health benefit from on-grid centralized photovoltaics ($0.09 million), on-grid distributed photovoltaics ($11.8 million), and off-grid photovoltaics ($28.7 million), was $40.6 million. Total benefits for 1976 to 2008 were $237 million.[12, 13]

Table 5.15 Environmental health benefits for on-grid centralized and on-grid distributed PV systems, 2008

Health effect	On-grid centralized		On-grid distributed	
	Incidence	Cost (2008$)	Incidence	Cost (2008$)
Mortality	0.01	82967	1.63	10875424
Infant mortality	<.01[a]	109	<.01	25638
Chronic bronchitis	0.01	3153	1.01	466212
Nonfatal heart attacks	0.02	1592	2.51	285379
Resp. hospital admissions	<.01	5	0.39	4211
CDV hospital admissions	0.01	59	0.81	21524
Acute bronchitis	0.02	0	2.42	484
Upper respiratory symptoms	0.19	0	21.6	229
Lower respiratory symptoms	0.26	0	28.63	168
Asthma ER visits	0.01	0	2.14	190
MRAD	10.57	382	1220.18	75679
Work-loss days	1.79	54	205.04	15583
Asthma exacerbations	0.25	–[b]	27.68	–
Total health effects		90495		11788589

Notes:

[a] Researchers have linked both short-term and long-term exposures to ambient levels of air pollution to increased risk of premature mortality. COBRA uses mortality risk estimates from an epidemiological study of the American Cancer Society cohort conducted by Pope et al. (2002). COBRA includes different mortality risk estimates for both adults and infants. Because of the high monetary value associated with prolonging life, mortality risk reduction is consistently the largest health endpoint valued in the study. COBRA rounds the incidence to zero from a very small value, but because the cost of mortality is high, even a very small value produces some cost.

[b] COBRA does not produce a value for asthma costs.

From 1976 to 2008, $39.8 million in environmental benefits can be attributed to DOE through gains in efficiency (Table 5.17).[14] Although total benefits were monetized using the COBRA model, specific attribution was unable to be resolved because of challenges associated with isolating technology effects from demand-side public policies. Thus, only a lower-bound estimate of environmental health benefit is presented. The exclusion of environmental health benefits has no material impact on the measures of economic return. Environmental health benefits were not included in the measures of economic return.

Table 5.16 Environmental health benefits for off-grid PV systems, 2008

Health effect	Incidence	Cost ($)
Mortality	3.95	26 335 489
Infant mortality	0.01	70 695
Chronic bronchitis	2.7	1 255 503
Nonfatal heart attacks	6.19	713 669
Resp. hospital admissions	0.92	11 938
CDV hospital admissions	1.91	54 230
Acute bronchitis	6.79	2 433
Upper respiratory symptoms	60.74	1 407
Lower respiratory symptoms	80.57	1 149
Asthma ER visits	3.25	787
MRAD	3 332.22	211 984
Work loss days	562.36	46 314
Asthma exacerbations	77.75	–[b]
Total health effects		28 718 032

Notes:

[a] Researchers have linked both short-term and long-term exposures to ambient levels of air pollution to increased risk of premature mortality. COBRA uses mortality risk estimates from an epidemiological study of the American Cancer Society cohort conducted by Pope et al. (2002). COBRA includes different mortality risk estimates for both adults and infants. Because of the high monetary value associated with prolonging life, mortality risk reduction is consistently the largest health endpoint valued in the study.

[b] COBRA does not produce a value for asthma costs.

ENERGY SECURITY BENEFITS

Solar energy represents a secure domestic source of energy in the face of threats to energy supply and provides clean energy to avoid long-run security risks from GHG emissions and climate change. Although national security benefits are difficult to monetize, they represent an important advantage of renewable energy. Because of its distributed nature, photovoltaics holds additional energy security benefits. In the US, 95 per cent of photovoltaics is distributed throughout small-scale on- and off-grid applications, making it less vulnerable to threats to the power supply than central power infrastructure.

Energy security benefits are presented quantitatively in barrel of oil equivalents (BOE). A BOE represents the energy released by burning a barrel of oil, or 1,700 kWh. The majority of on-grid photovoltaics provides energy that would normally be supplied by natural gas peaking plants, although some distributed photovoltaics replaces energy from coal and petroleum.

Table 5.17 Estimated environmental health benefits of PV attributable to DOE (2008$)

Year	Estimated baseline efficiency (%)	Estimated counterfactual efficiency (%)	Percentage difference	Total environmental health benefits ($ million)	Approximate benefits attributable to DOE ($ million)
1976	5.7	5.7	0%	0.1	0.0
1977	6.2	5.9	4%	0.1	0.0
1978	6.7	6.2	8%	0.2	0.0
1979	7.2	6.4	11%	0.3	0.0
1980	7.7	6.6	14%	0.5	0.1
1981	8.1	6.8	16%	0.7	0.1
1982	8.6	7.1	18%	0.9	0.2
1983	9.0	7.3	19%	1.1	0.2
1984	9.5	7.5	21%	1.3	0.3
1985	9.9	7.7	22%	1.6	0.4
1986	10.1	7.9	21%	1.8	0.4
1987	10.3	8.1	21%	1.9	0.4
1988	10.5	8.3	21%	1.9	0.4
1989	10.7	8.5	20%	2.0	0.4
1990	10.9	8.7	20%	2.0	0.4
1991	11.1	8.9	19%	2.1	0.4
1992	11.3	9.1	19%	2.5	0.5
1993	11.5	9.3	19%	2.9	0.5
1994	11.7	9.5	18%	3.3	0.6
1995	11.9	9.7	18%	3.8	0.7
1996	12.0	9.9	17%	4.5	0.8
1997	12.1	10.0	17%	5.2	0.9
1998	12.1	10.1	17%	6.0	1.0
1999	12.2	10.3	16%	7.0	1.1
2000	12.3	10.5	15%	8.3	1.2
2001	12.4	10.7	13%	9.7	1.3
2002	12.4	10.9	12%	11.7	1.4
2003	12.5	11.1	11%	14.2	1.6
2004	12.5	11.3	10%	18.1	1.8
2005	13.3	11.5	13%	21.7	2.9
2006	14.0	11.7	17%	26.2	4.3
2007	14.8	11.9	19%	32.7	6.3
2008	15.5	12.0	23%	40.6	9.2
Total				237.2	39.8

Source: Authors' calculations using COBRA.

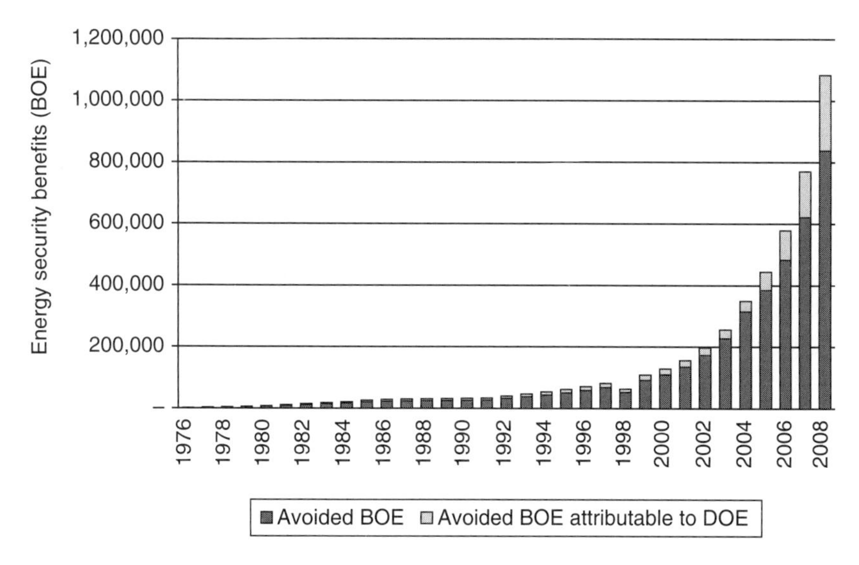

Source: Authors' calculations.

Figure 5.6 Energy security benefits (BOE)

Off-grid photovoltaics replaces internal combustion engines. In 2008, PV energy produced over 1.8 billion kWh, or 1.1 million BOE. Between 1976 and 2008, photovoltaics replaced an estimated 4.8 million BOE, of which approximately 0.8 million can be attributed to the DOE (Figure 5.6).[15]

CONCLUSIONS

This study quantified economic benefits of gains in reliability and reductions in cost attributable to the DOE from its long-term financial and technical support for PV module technology R&D. Our findings lead us to concur with, and provide measures of, economic and other benefits to substantiate, findings in technical reports, and policy studies that have concluded that the DOE has had a significant impact on the state of PV module technology.

Lower-bound measures of economic return were calculated for the DOE's investment in Photovoltaic Energy Systems by comparing quantified benefits accruing from a subset of funded technologies developed by private sector, university, and DOE researchers under FSA, PVMaT, and TFP. Between 1975 and 2008, the DOE invested $3,710 million (2008$) in Photovoltaic Energy Systems (Table 5.18). The total economic benefit accruing from this investment was $18,735 million, corresponding to

Table 5.18 Summary cost-benefit analysis results, 1975–2008

	Quantified benefit	Minimum attribution to DOE	Unit of measure
Economic benefits			
Net economic benefits	\$18734.8	\$18734.8	Million, 2008\$
Public rate of return		17%	
Net present value at 7% (Base year = 1975)		\$1458.9	Million, 2008\$
Net present value at 3% (Base year = 1975)		\$5724.7	Million, 2008\$
Benefit-to-cost ratio at 7%		1.83	
Benefit-to-cost ratio at 3%		3.24	
Environmental health benefits			
Monetized via COBRA	\$237.23	\$39.80	Million, 2008\$
Avoided mortality[a]	32.65	5.48	Deaths
Avoided infant mortality[a]	0.07	0.01	Deaths
Avoided chronic bronchitis	21.98	3.69	Cases
Avoided nonfatal heart attacks	51.03	8.57	Attacks
Avoided respiratory hospital admissions	7.63	1.28	Admissions
Avoided CDV hospital admissions	15.88	2.67	Admissions
Avoided acute bronchitis	54.87	9.20	Cases
Avoided upper respiratory symptoms	490.69	82.29	Episodes
Avoided lower respiratory symptoms	650.84	109.15	Episodes
Avoided asthma ER visits	29.52	4.99	Visits
Avoided MRAD	27036.52	4535.47	Incidences
Avoided work-loss days	685.87	123.00	Days
Greenhouse gas emissions benefits			
Avoided carbon dioxide emissions (CO_2)	6815103	1062473	Tons
Avoided methane emissions (CH_4)	132	21	Tons
Avoided nitrous oxide emissions (N_2O)	583	90	Tons

Table 5.18 (continued)

	Quantified benefit	Minimum attribution to DOE	Unit of measure
Avoided particulate matter emissions (PM)	1 232	207	Tons
Avoided sulfur dioxide emissions (SO_2)	2 634	463	Tons
Avoided ammonia emissions (NH_3)	16	3	Tons
Avoided volatile organic compounds emissions (VOCs)	1 090	181	Tons
Energy security benefits			
Equivalent avoided petroleum consumption	4 790 478	827 189	Barrels of oil equivalent
Knowledge benefits			
DOE-attributed patent families in photovoltaics		274	Patent families
DOE publications in photovoltaics		900	Publications
Percentage of leading US PV company patents linked to DOE		30%	

Notes: [a] Researchers have linked both short-term and long-term exposures to ambient levels of air pollution to increased risk of premature mortality. COBRA uses mortality risk estimates from an epidemiological study of the American Cancer Society cohort conducted by Pope et al. (2002). COBRA includes different mortality risk estimates for both adults and infants. Because of the high monetary value associated with prolonging life, mortality risk reduction is consistently the largest health endpoint valued in the study.

a return on DOE's investment of 17 per cent over the 33-year period. Applying a discount rate of 7 per cent yields a BCR of 1.83, indicating that for every $1 invested, $1.83 in benefits accrued. Applying a 3 per cent social discount increases the BCR to 3.24.

In addition to these economic benefits, other measures of benefit through 2008 were estimated:

- $237.2 million in environmental health benefits from avoided adverse health incidences, with approximately $39.8 million of these benefits attributable to the DOE.[16]

- 6.8 million tons of avoided CO_2 emissions, with approximately 1.1 million tons of avoided emissions attributable to the DOE.
- 4.8 million BOE in energy security benefits, with approximately 0.8 million of these attributable to the DOE.
- Knowledge benefits linking critical PV technology patents and publications at major US and international PV companies to DOE-funded or cost-shared R&D activities.

In addition to these quantitative measures, interviews with industry, academic, and public sector scientists and business leaders revealed that FSA, PVMaT, and TFP were critical to the development of PV companies. Experts concluded that without these programs not only would the state of photovoltaics be significantly poorer, but many US companies, which employ thousands of people, would not exist.

The influence of the DOE and the companies receiving cost shares is reflected in the scientific literature (factory automation for scale, encapsulants, thin-film photovoltaics, differential processing of ingots, measurement, and characterization) all this research was enabled by the DOE, which in turn reduced the LCOE and, in so doing, supported demand-side policies in fueling the accumulation of installed, clean, PV energy systems.

NOTES

1. These two studies are akin to engineering cost recovery analyses or payback analyses. They differ fundamentally from the current work in that they did not measure benefits relative to counterfactual technology or market development in the absence of the initiatives being reviewed. Witt et al. (2001) collected data on direct manufacturing cost per watt, production volumes, and production capacity from program participants. Year-on-year cost reductions were monitored, and program participants were asked to assign the proportion of annual cost savings passed to consumers via reduced prices or retained as increased profits. Cumulative cost savings estimates were compared with cumulative industry and public costs to gauge the timing of DOE investment recovery. The second study (Friedman et al., 2005) updated the results of Witt et al. (2001) with historical data from 2001 to 2005.
2. Friedman et al. (2005) adjusted dollar values to real terms using the consumer price index. This analysis corrected the inflation adjustment by reverting values back to nominal terms and then adjusting for inflation using the BEA national income production accounts.
3. Modules produced from 1976 to 1981, years in which no warranties existed, were assumed to have had a module lifetime of two years.
4. Production quantity included domestic and overseas production by companies receiving DOE cost share whose technology can be attributed directly to US-based R&D.
5. The US led the world in PV installations until surpassed by Japan in 1997 and Germany in 2001. Spain joined the top PV producers after increasing total installation by a factor of 23 between 2006 and 2008. The US had 1,169 MW of accumulated installations in

2008. Germany led with 5,340 MW, followed by Spain (3,354 MW) and Japan (2,144 MW) (IEA, 2009).
6. The average period of technology acceleration was the average of responses provided by researchers active between 1975 and 1990. Responses such as 'at least 10 years' or '10 to 15 years' were converted to lower-bound or midpoint estimates, respectively. Some experts were unable to provide an estimate but stated that the acceleration effect was 'significant' or 'fundamental.' The whole-year mean of responses was 12 years. In follow-up interviews, we reviewed with interviewees the effect their estimates would have on reliability milestones and costs.
7. Benefits were calculated for PV companies individually: Each firm's counterfactual cost per watt less actual cost per watt and then multiplied by their production volume yielded company-specific benefits. Benefits were then summed and assembled into a time series. Future analyses leveraging this work should take heed of the following. Accommodated within the analysis, but withheld from all tables are data on firm acceleration. Multiple companies indicated that they would not have existed in the absence of DOE funding, and others stated that not only were their cost-per-watt reductions accelerated, but their entire company's development was accelerated as well. Responses on company development acceleration ranged from 1 to 5 years. For the purpose of calculating economic benefits and assembling the counterfactual cost-per-watt curve, these companies' production quantity and costs were delayed by the acceleration period and remaining companies' data were used to create the curve. Thus, the difference between the counterfactual cost per watt and baseline curves differs from the implied average cost-per-watt benefit.
8. Furthermore, this analysis ignores economic benefits of the large-scale silane gas production at these facilities relative to alternative technologies and is, thus, a conservative estimate of the total benefit to society of this technology.
9. Chip fabrication machines require a standard-sized silicon wafer. The semiconductor industry has gone through a number of standard wafer diameters. Because both the machines to produce wafers and to fabricate chips are expensive, wafers of a particular diameter will persist for many years. Although wire saws are required for 300 mm production and the per-wafer savings are included in the total benefits, this analysis did not study acceleration of the progression to 300 mm adoption.
10. Following Ruegg and Jordan (2009), costs are assumed to be incurred at the beginning of each year, but benefits are assumed to be realized at the end of each year. Thus, the time period for the discounting of benefits is one year longer than for costs.
11. Avoided emissions estimated for 2008 have been scaled back to estimate previous years based on on-grid and off-grid kilowatt-hours production in each year. Because detailed data were not available for all years, this estimate assumes a constant ratio of distributed to centralized on-grid PV for the years prior to 2008. Including benefits for 2009 to 2033, assuming a useful life of 25 years, increases total GHGs avoided by 800 tons CH_4, 2,100 tons N_2O, and 32,100,000 tons CO_2. Thus, retrospective and future avoided GHGs for the installed base as of 2008 are 900 tons CH_4, 2,700 tons N_2O, and 38,900,000 tons CO_2.
12. Because of the linear relationship between benefits and kilowatt-hour generation, the benefits estimated for 2008 have been scaled back to estimate previous years based on kilowatt-hour production in each year. Because detailed data were not available for all years, this estimate assumes a constant ratio of distributed to centralized on-grid PV for the years prior to 2008 and a constant ratio of on-grid to off-grid PV before 1992.
13. Including benefits for 2009 to 2033, assuming a useful life of 25 years, increases total benefits before discounting by over $900 million. Thus, retrospective and future environmental benefits for the installed base as of 2008 are between $1.1 billion and $1.2 billion.
14. Including benefits projected for 2009 to 2033, approximately $270.3 million in environmental benefits can be attributed to DOE activities.
15. Including benefits for 2009 to 2033, assuming a useful life of 25 years, increases security

benefits by 24.9 million. Thus, retrospective and future benefits for the installed base as of 2008 are estimated at 29.7 million BOEs. An additional 5.7 million BOE can be attributed to DOE from the 2008 PV infrastructure extended out to 2033, amounting to a total of 6.5 million BOE in benefits.

16. Most photovoltaics in the US are installed in California, and environmental health and GHG emissions were compared with the likely next-best alternative energy portfolio. For California, this portfolio would likely consist of natural gas and other renewable energy sources. However, as electricity generation from PV installations in markets characterized by comparatively high coal combustion increases, such as in North Carolina and New Jersey, environmental benefits and avoided GHG emissions per kilowatt-hour would exceed those for California. Environmental health benefits were not included in the measures of economic return.

6. Investments in geothermal technologies

INTRODUCTION

Geothermal energy systems tap into hydrothermal energy in the earth to produce electricity. Geothermal energy has the advantage of being a clean, renewable energy source without the variability of other renewable sources, such as wind and solar. It is also a viable alternative for traditional fossil fuel (e.g., coal) base-load generation, particularly coal and natural gas.

Resources of geothermal energy vary in quality and accessibility due to differences in depth of reservoirs, rock formations, and water content. Historically, geothermal power plants have been built under ideal conditions for energy production. This is usually where the reservoir is close to the surface, the host rock is permeable and porous, and the ground fluid saturation and recharge rates allow having economically feasible operation. The relative scarcity of such ideal geothermal sites has been a barrier to widespread geothermal energy use (DOE, 2008c).

The DOE initiated the Geothermal Technologies Program (GTP) in the late 1970s and has conducted a wide range of research targeted at the long-term goal of making geothermal energy a cost-competitive power production alternative. For example, before research efforts by the GTP, little commercial geothermal power was generated in the US from the predominantly liquid-dominated hydrothermal resources.[1] Only four plants were installed from 1971 to 1979 (as compared with 16 plants from 1980 to 1985).

The US currently leads the world in online megawatt capacity of geothermal energy and electric power generation (Glitnir, 2008). However, the net electricity generated from geothermal power in the US in 2008 was 14,859 million kWh, or only 0.37 per cent of the total electricity generated in the US that year (DOE, 2010d). As shown in Table 6.1, the overwhelming majority of installed geothermal capacity is in California and Nevada (because of the abundance and ease of access to the heat sources in these states).

This chapter presents the findings from a retrospective economic analysis of the GTP conducted for the DOE (DOE, 2010c). A benefit-cost

Table 6.1 Geothermal power capacity by state, 2008

State	Installed capacity (MW)	Share by state (%)
California	2555.3	87
Nevada	318.0	11
Utah	36.0	1
Hawaii	35.0	1
Idaho	13.0	<1
Alaska	0.4	<1
New Mexico	0.2	<1
Total	2957.9	100

Source: Geothermal Energy Association (2009).

analysis was conducted for four selected technologies supported by the GTP. The objectives of the study were to

- Assess the DOE's role in technology development and adoption.
- Estimate the economic and environmental health benefits generated from selected technologies.
- Compare benefits attributable to the DOE's investments both for the GTP as a whole, and for the selected set of technologies examined in detail and estimate measures of economic return from the DOE's R&D activities.

Primary and secondary data collection was conducted for each of the technologies. As part of the study, 22 informal interviews were conducted with industry experts, academics, and DOE staff.

The economic and environmental benefit estimates associated with DOE activities are related primarily to lowering installation and operating costs, and increasing operating efficiencies and productivity. The exception is the DOE's impact on the adoption of binary cycle technologies, where the GTP's funding and demonstration projects proved the technology and accelerated its adoption.

Table 6.2 presents a summary of the benefits and GTP expenditures for each of the four technologies, along with the calculated measures of economic return. Following a discussion of the overall benefits and costs associated with each of the four selected technologies, additional detail is presented on the technology impact assessment methodology for PDC drill bits. Similar detailed analysis was conducted for the three other GTP technology areas included in the analysis and a complete description of this analysis can be found in DOE (2010c).

Table 6.2 Net benefits attributable to DOE (thousands, 2008$)

Year	PDC			Binary			TOUGH			Cement		
	Total benefits	Program expenses	Net benefits	Total benefits	Program expenses	Net benefits	Total benefits	Program expenses	Net benefits	Total benefits	Program expenses	Net benefits
1976	$0	$2081	−$2081	$0	$2036	−$2036	$0	$797	−$797	$0	$142	−$142
1977	$0	$2081	−$2081	$0	$2036	−$2036	$0	$797	−$797	$0	$142	−$142
1978	$0	$2081	−$2081	$0	$2036	−$2036	$0	$797	−$797	$0	$142	−$142
1979	$0	$2081	−$2081	$0	$2036	−$2036	$0	$797	−$797	$0	$142	−$142
1980	$0	$2081	−$2081	$0	$2036	−$2036	−$3158	$797	−$3955	$0	$142	−$142
1981	$0	$2081	−$2081	$0	$2036	−$2036	$7041	$797	$6244	$0	$142	−$142
1982	$319823	$2081	$317742	$0	$2036	−$2036	$3495	$797	$2698	$0	$142	−$142
1983	$400600	$2081	$398519	$0	$2036	−$2036	−$6065	$797	−$6862	$0	$142	−$142
1984	$630196	$2081	$628115	$2114	$2036	$78	$10446	$797	$9649	$0	$142	−$142
1985	$662306	$2081	$660225	$2047	$2036	$11	$6455	$797	$5658	$0	$142	−$142
1986	$436938	$2081	$434857	$9822	$2036	$7786	$21896	$797	$21099	$0	$142	−$142
1987	$443817	$1844	$441973	$14398	$1639	$12759	$21946	$797	$21149	$0	$142	−$142
1988	$474236	$1683	$472553	$3750	$1335	$2415	$15404	$797	$14607	$0	$142	−$142
1989	$465778	$1601	$464177	$8238	$1015	$7223	$28656	$797	$27859	$0	$142	−$142
1990	$619142	$1490	$617652	$15229	$1025	$14204	$38331	$797	$37534	$0	$142	−$142
1991	$610952	$1586	$609366	$7337	$2277	$5060	$44331	$797	$43534	$0	$142	−$142
1992	$548896	$1575	$547321	$15926	$3542	$12384	$42944	$159	$42785	$0	$142	−$142
1993	$655894	$1455	$654439	$26175	$3745	$22430	$44169	$159	$44010	$0	$142	−$142
1994	$640427	$1335	$639092	$6700	$3770	$2930	$42213	$159	$42054	$0	$142	−$142
1995	$618012	$1341	$616671	−$12363	$2890	−$15253	$34046	$159	$33887	$0	$142	−$142
1996	$731344	$1474	$729870	−$5514	$1844	−$7358	$35623	$159	$35464	$0	$142	−$142
1997	$922792	$1558	$921234	$3362	$1925	$1437	$38239	$159	$38080	$0	$142	−$142
1998	$778133	$1817	$776316	$978	$1787	−$809	$37958	$159	$37799	$0	$142	−$142
1999	$564958	$1979	$562979	−$391	$1616	−$2007	$37664	$159	$37505	$41	$142	−$101
2000	$862685	$2045	$860640	−$2399	$1452	−$3851	$33581	$159	$33422	−$123	$142	−$265

Table 6.2 (continued)

Year	PDC			Binary			TOUGH			Cement		
	Total benefits	Program expenses	Net benefits	Total benefits	Program expenses	Net benefits	Total benefits	Program expenses	Net benefits	Total benefits	Program expenses	Net benefits
2001	$1 454 710	$2 188	$1 452 522	−$2 202	$1 203	−$3 405	$35 559	$159	$35 400	$1	$150	−$149
2002	$1 454 567	$2 188	$1 452 379	$2 226	$1 040	$1 186	$37 844	$159	$37 685	$165	$150	$15
2003	$2 025 308	$2 188	$2 023 120	$2 489	$971	$1 518	$37 994	$159	$37 835	$291	$150	$141
2004	$2 497 349	$2 188	$2 495 161	$1 721	$953	$768	$39 428	$159	$39 269	$472	$120	$352
2005	$3 429 701	$2 188	$3 427 513	$6 471	$972	$5 499	$38 802	$159	$38 643	$1 540	$130	$1 410
2006	$4 631 182	$2 188	$4 628 994	$4 882	$972	$3 910	$38 797	$159	$38 638	$1 666	$150	$1 516
2007	$5 589 616	$2 188	$5 587 428	$2 677	$972	$1 705	$40 735	$159	$40 576	$1 811	$142	$1 669
2008	$6 081 340	$2 188	$6 079 152	$10 533	$972	$9 561	$41 817	$159	$41 658	$1 955	$142	$1 813
Undis-counted total	$38 550 702	$63 178	$38 487 524	$124 206	$60 313	$63 893	$846 193	$15 455	$830 738	$7 820	$4 684	$3 136
PV[a] at 7%	$7 813 212	$26 461	$7 786 751	$42 848	$26 819	$16 029	$219 445	$8 619	$210 826	$1 013	$1 938	−$925
PV[a] at 3%	$18 514 201	$41 015	$18 473 186	$76 269	$40 701	$35 568	$457 957	$11 655	$446 302	$3 199	$3 037	$162
BCR at 7%	295.3			1.6			25.5			0.5		
BCR at 3%	451.4			1.9			39.3			1.1		
IRR	139%			15%			48%			NA		

Note: [a] PV base year is 1976.

Individual Technology Case Studies

Polycrystalline diamond compact drill bits

Technology description In the early 1970s, the Geothermal Division of the DOE began research to produce an enhanced drill bit that would be more suitable than traditional drill bits for the high-density, high-temperature applications needed to drill geothermal wells (EERE/GTP, 2008a). This led to the development of several technological drilling advances, among which were PDC drill bits. A typical PDC drill bit consists of the drill bit body, usually made from steel or matrix metal, and drill bit cutters, their number varying with the bit diameter. PDC drill bit cutters are made from synthetic diamond powder (heated and pressurized graphite), cobalt, and tungsten carbide, and the bits have the key advantage of having no moving parts.

Next best alternative PDC drill bit technology is an improvement to an existing technology. The next-best alternative technology is the traditional moving parts roller-cone bit. Roller-cone bits are an established technology and continue to be used in applications involving shallow wells and softer rock formations. With its harder and longer-lasting cutting surface, the PDC bit uses a simpler mechanical action, increasing productivity (feet drilled per hour) and efficiency (decreasing the number of drill bits per well).

Economic benefits Approximately 60 per cent of worldwide oil and gas well footage in 2006 was drilled using PDC drill bits (Blankenship, 2009). The main advantage of PDC drill bits over conventional roller cone bits[2] is that they reduce frequency of pulling the drill string to replace the drill bit, allowing higher penetration rates and thus reducing the time (and cost) of renting expensive drill rigs. The use of PDC drill bits in offshore applications in the oil and gas industry is estimated to reduce costs by $59 per foot drilled, yielding a present value cost savings of $15.6 billion[3] from 1982 to 2008.

Environmental benefits Using PDC drill bits does not directly affect emissions. Environmental impacts were judged to be small and were not quantified.

Security benefits Using PDC drill bits reduces the cost and increases the availability of domestic oil and gas, which can reduce US dependence on foreign imports of oil and gas. However, this potential impact was not quantified.

DOE attribution The DOE played a very important role in developing and adopting the PDC drill bit technology, making significant contribution to (1) developing the bit and getting it to the market, (2) overcoming performance flaws, and limitations and (3) spurring the innovation that resulted in overall market success of PDC drill bits. As a result, based on the observable technology transfer and findings from published papers and interviews, this study attributes 50 per cent of the economic benefits from PDC bits to the DOE. This attribution estimate is consistent with previous analysis (see Papadakis and Link [1997] and Falcone and Bjornstad [2005]). Thus, the present value of total benefits from the DOE's contributions to developing drill bit technology equals $7.8 billion.

Measures of economic return DOE program expenditures associated with PDC drill bit research are presented in Table 6.2 and total $26.5 million from 1982 to 2008. Comparing the DOE's investment in PDC drill bit technology with benefits attributed to the DOE yields a net benefit with a present value of $7.8 billion ($18.5 billion discounted at 3 per cent). The BCR is calculated to be 295.3 discounted at 7 per cent (451.4 discounted at 3 per cent), and the IRR for the PDC drill bit technology project was 139 per cent.

Binary cycle power plant technology

Technology description Binary cycle plants represented 14.5 per cent of total geothermal capacity in the US in 2008. Binary cycle geothermal power plant technology enables efficient use of lower-temperature resources through the use of a closed-loop heat transfer system. The DOE assisted with the penetration of binary cycle plant technology in the US, by sponsoring research to lower costs and by developing demonstration projects to showcase the technology's viability.

Next best alternative The next-best alternative to binary cycle is usually flash cycle technology, which was used in 84 per cent of geothermal power plants in 2008 (GEA, 2009). However, the appropriate counterfactual technology for the benefits analysis differs depending on the temperature of the reservoir where binary cycle is installed:

- In reservoirs where the temperature range is 150°C to 190°C, flash cycle technology is economically viable but has lower electricity generation productivity compared with binary cycle because of its lower conversion efficiency. Thus, in this temperature range, the next-best alternative is a traditional, but less productive, flash cycle geothermal plant.

Table 6.3 Security benefits of binary cycle attributable to DOE, 1980–2008

Fossil fuel	Binary cycle
Natural gas (million cubic feet)	118207
BOE (thousand barrels of oil equivalent)	20951

- In reservoirs where the temperature range is below 150°C, flash cycle technology is not economically viable, and no other geothermal technology is available. Hence, for temperature ranges below 150°C, the next-best alternative technology is generation from a mix of fossil fuel.

Economic benefits Economic benefits include the market value of the additional electricity produced from reservoirs where the temperature range is 150°C to 190°C.[4] The present value of economic benefit is estimated to be approximately $109.7 million (discounted at 7 per cent) from 1984 to 2008.

Environmental benefits Environmental benefits result from both (1) the additional renewable electricity generated from reservoirs where the temperature range is below 150°C, offsetting generation from a mix of fossil fuel and (2) the incremental renewable power generation (which has zero emissions) from reservoirs where the temperature range is 150°C to 190°C, which would have been flash cycle-generated electricity but are now binary cycle-generated electricity. The present value of environmental health benefits is estimated to be $127.0 million (discounted at 7 per cent). In addition there are reductions of GHGs totaling 7.3 million tCO_2e from 1984 to 2008.

Security benefits Security benefits derive from reducing the probability and potential impact of oil and natural gas disruptions and price shocks or other energy system disruptions that would damage or disrupt the economy, environment, or national security of the US. Table 6.3 presents the reduction in 4.9 billion cubic feet of natural gas or 862000 million BOE.

DOE attribution The main impact of the DOE on binary cycle technology has been the demonstration of commercial applicability of binary cycle technology and the provision of guaranteed loans, which helped industry obtain financing. This accelerated the entry of the technology into the market. Interview participants estimated that the acceleration effect was less than five years but more than six months. Based on that estimate, it was assumed that the acceleration effect of DOE activities was two years. The present value of benefits attributable to the DOE's acceleration effect equaled approximately $42.8 million (discounted at 7 per cent).

Measures of economic return DOE expenditures on binary cycle activities represent approximately 1.6 per cent of EERE's total GTP budget from 1976 to 2008. The present value of expenditures (adjusted to 2008$) equaled $26.8 million discounted at 7 per cent ($41 million discounted at 3 per cent). When expenditures are compared with the benefits with present value of $42.8 million discounted at 7 per cent attributed to the DOE, the DOE's investment in binary cycle power plant technology yielded net benefits with the present value of $16.0 million discounted at 7 per cent ($35.6 million discounted at 3 per cent). The BCR discounted at 7 per cent is calculated to be 1.6 (1.9 discounted at 3 per cent), and the IRR for binary cycle plant technology is 16 per cent.

TOUGH series of reservoir models

Technology description Reservoir models simulate the flow of fluids and thermal energy in the subsurface. They are used as analytical and management tools in forecasting reservoir capacity, designing development activities, and identifying characteristics that may cause changes in the reservoir. Reservoir modeling helps optimize resources and manage risk. The DOE's research efforts helped identify the limitations of traditional exploration techniques and demonstrated the potential for imaging and characterizing subsurfaces, which provided the foundation for geothermal reservoir modeling.

Next-best alternative Geothermal reservoir modeling, in general, is an example of new technology representing analytic capabilities of much higher complexity than previously available. Before capabilities for detailed computer simulation of geothermal reservoirs were developed, the best alternative was to use so-called 'lumped parameter' models. These models simplified reservoir simulation and provided limited information about how fast this energy could be recovered, how many production and injection wells would be required, where these wells should be located, and at what rate makeup wells would be required as the reservoir was being depleted. This sometimes led to incorrect estimation of power-plant capacity, thus increasing installed costs.

The benefits associated with using reservoir modeling that were quantified for this study fall into two categories: (1) increased productivity of geothermal resources (the value of the additional electricity generated) and (2) the environmental health benefits associated with the additional renewable energy offsetting fossil fuel generation.

Table 6.4 Security benefits of TOUGH models attributable to DOE, 1980–2008

Fossil fuel	TOUGH models
Natural gas (million cubic feet)	19401
BOE (thousand barrels of oil equivalent)	3438

Economic benefits Using reservoir modeling increased productivity of geothermal resources. These benefits are somewhat offset by additional exploration costs associated with reservoir modeling. However, in the aggregate, reservoir modeling has been profitable for the geothermal industry by improving subsurface exploration. The present value of total economic benefits of reservoir modeling from 1984 to 2008 equaled $503.3 million (discounted at 7 per cent). In addition, reservoir modeling has been adapted for other uses, such as nuclear waste geologic storage, CO_2 sequestration, and remediation of subsurface contamination problems. For these applications, the TOUGH model has been the primary modeling tool; however, this study was not able to quantity the benefits of nongeothermal applications.

Environmental benefits The increased productivity of geothermal fields yields additional renewable energy that offsets generation from a mix of fossil fuel. The present value of the environmental health benefits associated with the additional renewable energy is estimated to be $450.8 million (discounted at 7 per cent). In addition, there are reductions of GHGs totaling 22.9 million tCO_2e.

Security benefits Security benefits derive from reducing the probability and potential impact of oil and natural gas disruptions and price shocks or other energy system disruptions that would damage or disrupt the economy, environment, or national security of the US. Table 6.4 presents the reduction in 19.4 billion cubic feet of natural gas or 3 million BOE.

DOE attribution The DOE's activities in the late 1970s and early 1980s were at the forefront of geothermal reservoir model development, coinciding with the evolution of computer capabilities that enabled complex simulation models to be run cost-effectively. To quantify attribution, this study partitioned the suite of available reservoir models into two groups: (1) the TOUGH series of models and (2) other reservoir models (e.g., TETRAD, STAR). Participants interviewed as part of this study indicated that the DOE attribution between the two groups is different:

- The DOE had overwhelming influence (80 per cent) on the TOUGH series models.
- The DOE efforts were influential (20 per cent) on other reservoir models.

Based on a study conducted by O'Sullivan et al. (2001), the share of TOUGH usage for US geothermal applications is approximately 5 per cent, with other reservoir models accounting for 95 per cent of US geothermal reservoir modeling. Taking these usage shares and the attribution rates, this study estimates that the present value of total benefits discounted at 7 per cent attributable to the DOE equaled $219.4 million for the period 1980 to 2008.

Measures of economic return The present value of DOE program expenditures, discounted at 7 per cent, associated with the TOUGH family of reservoir models (adjusted to 2008$) equaled $8.6 million from 1976 to 2008 ($11.7 million discounted at 3 per cent). When compared with attributed benefits of $219.4 million discounted at 7 per cent, the DOE's investment in the TOUGH models yielded net benefits with a present value of $210.8 million discounted at 7 per cent ($446.3 million discounted at 3 per cent). The BCR is calculated to be 25.5 discounted at 7 per cent (39.3 discounted at 3 per cent). The IRR for the TOUGH program is 48 per cent.

High-temperature geothermal well cements

Technology description Early geothermal energy projects revealed that using Portland cement in geothermal wells was problematic, leading to frequent and costly repairs and significantly shorter well lifetimes. In the 1970s, the DOE began supporting the development of new well cements that address these shortcomings through basic materials research and applied research on the cementitious properties of various chemical formulations. DOE's research led to the patenting and commercialization of a calcium aluminate phosphate (CaP) cement system that is resistant to acidic corrosion and maintains structural integrity at extremely high temperatures.

Next-best alternative High-temperature well cement is a technology improvement over existing Portland-based well cements commonly used in geothermal, oil and gas wells. Originally developed for use in geothermal wells, high-temperature cement has also been used for enhanced oil recovery projects and offshore well drilling. The DOE-developed cement technology affects new well construction and ongoing maintenance at high-temperature geothermal production wells.

Economic benefit The rapid deterioration of Portland cement in geothermal wells (<12 months) resulted in frequent well workovers and costly well remediation. The use of high-temperature cements enhances performance in terms of structural stability and corrosion resistance, and is estimated to eliminate $150,000 in annual well remediation costs as well as extend the working life of geothermal production wells to 20 years or more. The economic benefits from high-temperature cements include (1) cost savings to the end-users of CaP cement and (2) profits from the sale of ThermaLock cement. The present value of total benefits from the use of high-temperature cement from 1999 to 2008 is estimated to be $2 million (discounted at 7 per cent), with 99 per cent of the benefits associated with cost savings to users.

Environmental benefits No environmental benefits associated with using high-temperature cement were identified. However, a complete life-cycle analysis including cement production (beyond the scope of this study) might identify emission reductions in GHGs and other pollutants.

Security benefits No security benefits associated with using high-temperature cement were identified.

DOE attribution The DOE's influence on the development of high-temperature cement varied over the 24-year period examined. For example, the DOE had a very important influence on determining the direction of cement research, choosing to pursue a ceramic-based cement formulation over a more conventional Portland-based design. In contrast, since commercialization in 1999, Halliburton has marketed ThermaLock for use in domestic and international geothermal and enhanced oil recovery injection wells with minimal direct involvement from DOE. Averaging estimated influence factors over each stage in the technology development cycle yields an estimated 48 per cent attribution rate. Thus, roughly half of the total benefits realized from the development of CaP cement, or approximately $1 million, are directly attributable to GTP's R&D activities.

Measures of economic return The present value of the DOE program expenditures associated with cement materials research (adjusted to 2008$) equaled $1.9 million (discounted at 7 per cent) from 1976 to 2008 ($3 million discounted at 3 per cent). Compared with the economic benefits attributed to the DOE, the DOE's investment in CaP cement technology yielded a present value of -$925 million discounted at 7 per cent in net benefits ($162 million discounted at 3 per cent). However, it is important to note that the technology is in its infancy, and experts interviewed for this analysis predict a considerable increase in the rate of adoption over the next five years. Allowing more time for industry to adopt the existing CaP cement technology in geothermal wells would significantly increase the total economic benefits realized and alter the results of this analysis. Depending on the rate of adoption, economic benefits may potentially exceed the total development costs, yielding a positive return to program costs in the near term.

BENEFIT-COST ANALYSIS FOR THE GTP

The analysis of economic, health, and knowledge benefits is summarized in this section.

Quantified Economic and Environmental Health Benefits

The four technologies selected for analysis reflect the wide range of research activities conducted by the GTP and, as a group, have generated significant economic and environmental benefits. Table 6.5 shows the aggregate monetized benefits for the four technologies and partitions them into general categories. Cost reductions in drilling and exploration accounted for the majority of the quantified benefits with a present value of $7.8 billion discounted at 7 per cent. A present value of increased operating efficiency and productivity accounted for $135.6 million discounted at 7 per cent and environmental health impacts accounted for $126.6 million discounted at 7 per cent.

In addition, as shown in Table 6.6, research activities associated with these four technologies accounted for only 3.8 per cent of GTP's budget from 1976 to 2008 (based on cost data discounted at 7 per cent). Hence, it is very likely that the total benefits associated with GTP activities greatly exceed the $8.1 billion quantified from the four selected technologies. However, it is still informative to compare the benefits from the four

Table 6.5 Summary of GTP benefits, 1976–2008 (thousands, 2008$)

Benefits	Total benefits PV[a] at 7% ($)
Cost reduction in drilling and exploration	7814225
Increased operating efficiency and productivity	135649
Environmental health impacts	126644
Total benefits attributable to GTP	8076518

Notes: [a] PV base year is 1976.

Table 6.6 Summary of GTP expenditures, 1976–2008

Technologies	Program expenses PV[a] at 7% (thousands, 2008$)	Percentage of total
PDC drill bits	26463	1.6
High-temperature cement	1934	0.1
TOUGH models	8620	0.5
Binary cycle plants	26819	1.6
Total of four technologies	63836	3.8
Total GTP expenditures	1660194	100.0

Notes: [a] PV base year is 1976.

technologies to the total GTP expenditures to obtain lower-bound measures of economic return.

Table 6.7 presents the cluster analysis results for the GTP. The present value of net benefits (in 2008$) for the program is $6.4 billion discounted at 7 per cent ($17.0 billion discounted at 3 per cent). The BCR is 4.8 discounted at 7 per cent (8.9 discounted at 3 per cent). The IRR is 22 per cent.

PDC DRILL BIT TECHNOLOGY: DETAILED IMPACT ASSESSMENT

This section provides additional detail on the technology impact assessment for PDC drill bits. Similar analysis was conducted for the three other technology areas included in the analysis and a complete description of this analysis can be found in DOE (2010c).

Approximately 60 per cent of worldwide oil and gas well footage in 2006 was drilled using PDC drill bits. A typical PDC compact drill bit consists of the drill bit body, usually made from steel or matrix metal, with synthetic diamond (PDC) cutters attached. Figure 6.1 compares typical PDC drill bits to roller-cone drill bits. DOE initially sponsored PDC drill bits as a potential application for geothermal wells and they have since been widely adopted by the oil and gas industry. This section discusses the history and social benefits of PDC drill bit technology attributable to the DOE, and the DOE's role in bringing this technology to the US market. The present value of total benefits related to PDC drill bit technology is estimated to be approximately $15 billion (discounted at 7 per cent). Approximately $7.8 billion (discounted at 7 per cent) of these benefits can be attributed to the DOE through its research activities resulting from an investment with a present value of $26.5 million discounted at 7 per cent.

Table 6.7 Benefit-cost analysis for GTP, 1976–2008 (thousands, 2008$)

Year	Geothermal Program		
	Total benefits	Program expenses[a]	Net benefits
1976	$0	$92819	−$92819
1977	$0	$130899	−$130899
1978	$0	$288654	−$288654
1979	$0	$367328	−$367328
1980	−$3158	$316935	−$320093
1981	$7041	$324090	−$317049
1982	$323318	$142327	$180991
1983	$394536	$93412	$301124
1984	$642756	$41677	$601079
1985	$670808	$57793	$613015
1986	$468656	$45668	$422988
1987	$480161	$34891	$445270
1988	$493391	$35623	$457768
1989	$502671	$30829	$471842
1990	$672701	$26832	$645869
1991	$662620	$43236	$619384
1992	$607766	$38143	$569623
1993	$726238	$31619	$694619
1994	$689340	$31251	$658089
1995	$639695	$50360	$589335
1996	$761452	$38384	$723068
1997	$964393	$37373	$927020
1998	$817069	$36402	$780667
1999	$602273	$35194	$567079
2000	$893744	$28554	$865190
2001	$1488069	$32205	$1455864
2002	$1494803	$32149	$1462654
2003	$2066083	$30550	$2035533
2004	$2538971	$29348	$2509623
2005	$3476513	$27414	$3449099
2006	$4676527	$24478	$4652049
2007	$5634838	$5107	$5629731
2008	$6135647	$19307	$6116340
Undiscounted total	$39528921	$2600851	$36928070
PV[b] at 7%	$8076518	$1660194	$6416324
PV[b] at 3%	$19051625	$2082623	$16969002
BCR at 7%			4.9
BCR at 3%			9.1
IRR			22%

Source:

[a] DOE (2008b).

[b] Base year is 1976, which is the first year of DOE program expenses.

PDC drill bits

Roller-cone drill bits

Source: EERE/GTP (2008c).

Figure 6.1 PDC and roller-cone drill bits

History of the PDC Technology

In the early 1970s, OPEC and individual oil-producing countries instituted higher prices and oil embargoes from 1973 to 1985. This put pressure on the US oil companies to expand oil production to more challenging geographic regions, such as deeper wells in the Gulf of Mexico (Jones, 1988), and to look for other energy sources. These events led to the development of several technological drilling advances, among which was the PDC drill bit. At that time, government labs began research to produce an enhanced drill bit that would reduce drilling costs for geothermal wells. Drilling is a large part of the capital cost of a geothermal power plant; thus, cheaper drilling provides a definite stimulus for placing geothermal power online.

High temperatures in geothermal wells cause serious damage to traditional roller-cone bits, partly because of damage to the bearing seals and bearings. PDC drill bits have the advantage of no moving parts. A typical PDC drill bit consists of the drill bit body, usually made from steel or matrix metal (Mensa-Wilmot, 2003), with PDC cutters attached. The number of cutters varies with the bit diameter and the hardness of the formation that it is designed to drill. The cutters are made under immense pressure and temperature from synthetic diamond powder sintered with cobalt onto a tungsten carbide substrate. The resulting disks (also known as compacts) can be made in different sizes but are typically half an inch in diameter and 0.3 of an inch thick. The disks are bonded to tungsten carbide cylinders to form cutters, and the cylinders are then brazed into the bit body (Falcone and Bjornstad, 2005).

General Electric developed synthetic diamonds in 1955 and first used them on prototype drill bits in the field in 1973 (Madigan and Caldwell,

1981). However, field versions of early bits were disappointing, because the compacts detached from their mounts and wore out quickly. To address these issues, beginning in the 1970s, SNL conducted in-house research and promoted industry R&D by funding field tests and fundamental studies of rock-cutting interactions and frictional heating of the cutters. The research was broadly focused on the following areas (Finger and Glowka, 1989):

- Bit-rock interaction to understand how cutters induce failure in rock.
- Diffusion bonding to prevent cutters from detaching from the mounting studs or the bit body.
- Cutter temperature modeling to understand how frictional heating affects the wear behavior of PDC cutters.
- Single-cutter tests to model bit behavior as a function of a combination of parameters, such as vertical force, depth of cut, type of rock, rake angle, and lubricity of the drilling fluid.
- Bit design modeling to determine the layout of cutter patterns on a PDC

The first commercial application of PDC drill bits occurred in 1976, and in 1977, General Electric marketed the first PDC cutter under the name Stratapax (Slack and Wood, 1982). PDC bits were initially less successful in geothermal formations because the principal wear mechanism of PDC drill bits is frictional heating. High heat, and hard and fractured rock formations, aggravated that problem. Nonetheless, PDC bits drill soft formations, at high temperatures. Successful PDC drill bit research was adopted by the oil and gas industry.

Cooperation between the oil and gas industry and SNL began during the period 1973 to 1977, while the lab was working with General Electric on improving the performance of the PDC drill bit. SNL sponsored wear and friction tests and conducted research on drill mechanics and hydraulics. After drill bit commercialization in 1977, Sandia kept working with the industry (because it was willing to participate in research that was mutually beneficial to the geothermal program and oil and gas industry), and the R&D program continued (Papadakis and Link, 1997). SNL helped resolve several technical problems exhibited by PDC bits and aided in bit design. SNL's computer program, STRATAPAX (released in 1982), and later PDCWEAR (released in 1986), helped manufacturers place cutters on the bit strategically (Falcone and Bjornstad, 2005). PDCWEAR could 'compare bit designs and gain detailed information on the individual cutters so that the bit design can optimally place the cutters to produce uniform cutter wear. PDCWEAR also predicted the performance of specific bit-rock combinations' (Finger and Glowka, 1989, p. 63).

SNL's PDC R&D program ran through 1986. After this, SNL's scientists and engineers continued to contract with a consortium of PDC drill bit manufacturers and university researchers to work on advancing the PDC technology (Glowka and Schaefer, 1993). For example, in 1995, Sandia's scientists collaborated with five drill bit companies (in the research sponsored by the industry) to improve the performance of PDC drill bits in harder rock formations (Glowka et al., 1995), and in 2004 SNL cooperated with bit manufacturers in testing 'best effort' drag bits in extensive drilling tests (Wise et al., 2004).

In general, advances in PDC drill bit technology since the first commercial application can be broken into three overlapping time segments:

- From 1977 to 1986, fundamental design and manufacturing deficiencies were identified and corrected.
- From 1979 to 1998, enabling research in bit mechanics, hydraulics, and thermal effects was performed. The PDCWEAR computer code was developed, which enabled the development of anti-whirl drill bits. Also, best practices in operation were established during this period.
- From 1996 to 2006, drilling dynamics were addressed, and improved cutter structures were tested and implemented. Integrated bit/bottom hole assembly was developed, and rotary steerable subs were introduced (Blankenship, 2009).

Experts and practitioners agree that SNL played an important role in developing PDC drill bit technology. The lab's contribution to PDC technology included providing financing and R&D contracts to General Electric to run wear and friction tests, performing internal research on PDC drill mechanics and hydraulics, performing PDC drill bit tests in the field, and resolving technical problems with PDC drill bits (Papadakis and Link, 1997). The transfer of knowledge from the DOE to drill bit manufacturers occurred through peer-to-peer discussion and numerous presentations and papers published by SNL.

Next-Best Alternative for PDC

PDC drill bit technology is an example of improvement to an existing technology. The next-best alternative technology is the traditional moving parts roller-cone bit. Roller-cone bits are an established technology and continue to be used in applications where PDC bits are unsuitable, such as very hard formations. With its harder and longer-lasting cutting surface, the PDC bit uses a more efficient mechanical action, shearing rock instead of crushing it.

This mechanical action increases productivity (feet drilled per hour) and reduces the frequency of pulling the drill string to exchange a drill bit, thus increasing efficiency (decreasing the number of drill bits per well). Compared with roller-cone drill bits, PDC drill bits reduce the time and costs of drilling. Both PDC and roller-cone drill bits have improved over time. However, there is limited empirical data to measure improvements in cost-effectiveness of both drill bit types. Thus, this study uses fixed 'delta' technical impact metrics over the time period of this analysis. This assumes that improvements over the technology's lifespan are comparable for both the PDC and the alternative roller-cone drill bits.

PDC Benefits Calculations

Benefits realized from using PDC drill bit technology are primarily economic benefits and are segmented into profits to manufacturers of PDC drill bits and cost reductions to oil producers from using PDC drill bits in oil exploration. Table 6.8 summarizes the key parameters and assumptions used to estimate benefits.

Table 6.8 Key parameters and assumptions used in the PDC drill bit benefits analysis

Parameters/Assumptions	Source
PDC drill bit market penetration 60% by 2008	Blankenship (2009)
Crude oil and natural gas exploratory and developmental well footage drilled	Crude Oil Developmental and Exploratory Well Footage (DOE, 2010a); Natural Gas Developmental and Exploratory Well Footage (DOE, 2010b)
PDC drill bits yield a cost reduction of $59 per foot drilled	Average of seven empirical studies published between 1982 and 1997
6.5% net profit estimate for drill bit producers	Falcone and Bjornstad (2005)
50% attribution of benefits to DOE	Published literature including Papadakis and Link (1997), interviews with industry and DOE experts

Economic benefits (profits) to PDC drill bit producers

To estimate the benefits to PDC drill bit producers, historical PDC drill bit sales were studied, starting in 1982, which was the first year for which the PDC market could be defined (Falcone and Bjornstad, 2005). This study used data from three sources to construct sales estimates from 1982 to 2008: Falcone and Bjornstad (2005) provided sales estimates for 1982 to 1992, DOE (2010c) provided sales estimates for 2000, and Freedonia Group (2009) supplied sales estimates for 2007. Sales for 1993 to 1999 and 2001 to 2008 were calculated from the above data using linear interpolation. The estimated sales for the entire period are presented in Table 6.9.

We considered profit margins for the four largest PDC drill bit manufacturers: Baker Hughes, Smith Bits, ReedHycalog, and Security-DBS. Unfortunately, all four top producers are subsidiaries of large oil (service or tool and equipment) companies: Baker Hughes of Baker International, Smith Bits of Smith International, Security-DBS of Halliburton, and ReedHycalog of National Oilwell Varco. Parent companies did not provide profitability numbers for subsidiaries, and using parent company profitability information would be misleading. Falcone and Bjornstad (2005) estimated 6.5 per cent as a net profit estimate for drill bit producers. Therefore, 6.5 per cent was used as a net profit estimate for drilling companies. The profit calculation for drill bit manufacturers is presented in Table 6.9.

Economic benefits (cost reductions) from using PDC drill bits

One of the main expenditures to oil production companies is the cost of renting drilling rigs. Regardless of whether the well is successful, drilling companies could pay up to \$1,000,000 (in 2008\$) for a drilling rig per day (Falcone and Bjornstad, 2005; Blankenship, 2009). Oil producers often use cost per foot calculations as a measure of drilling efficiency. The following formula is used to calculate cost per foot:

$$\frac{\text{Cost}}{\text{Foot}} = \frac{[(\text{Drilling Time} + \text{Trip Time}) \times \text{Hour Rig Cost} + \text{Bit Cost} + \text{Tool Cost} + \text{Labor Cost}]}{\text{Footage Drilled}}$$

The main advantage of PDC drill bits over the conventional roller-cone bits is that they allow higher penetration rates and reduce the frequency of changing the drill bit, Thus reducing the time of renting expensive drill rigs. Even though PDC bits cost more than roller-cone bits, they produce net benefits of cost per foot. To calculate the benefits to the oil extraction industry resulting from cost savings of drilling with PDC bits, this study used the following formula:

Table 6.9 Profits to PDC drill bit manufacturers from PDC drill bit technology, 1982–2009

Year	Annual sales of PDC drill bits (thousands $current)	Annual sales of PDC drill bits (thousands $2008)	Profits[a] (thousands $2008)
1982	$300 697	$588 679	$38 264
1983	$377 503	$710 928	$46 210
1984	$595 101	$1 080 234	$70 215
1985	$626 449	$1 103 680	$71 739
1986	$413 769	$713 149	$46 355
1987	$420 911	$705 044	$45 828
1988	$450 533	$729 608	$47 425
1989	$443 303	$691 796	$44 967
1990	$590 329	$886 913	$57 649
1991	$583 451	$846 563	$55 027
1992	$524 734	$743 776	$48 345
1993	$627 620	$870 365	$56 574
1994	$613 365	$833 037	$54 147
1995	$592 408	$788 196	$51 233
1996	$701 587	$916 030	$59 542
1997	$885 875	$1 136 466	$73 870
1998	$747 337	$948 036	$61 622
1999	$542 911	$678 724	$44 117
2000	$829 704	$1 015 301	$65 995
2001	$1 400 276	$1 675 773	$108 925
2002	$1 400 979	$1 649 763	$107 235
2003	$1 952 203	$2 250 638	$146 291
2004	$2 409 614	$2 701 058	$175 569
2005	$3 312 964	$3 594 016	$233 611
2006	$4 478 373	$4 704 668	$305 803
2007	$5 410 144	$5 525 630	$359 166
2008	$5 890 035	$5 890 035	$382 852
Undiscounted total			$2 858 577
PV[b] at 7%			$602 767

Notes:
[a] Calculations are based on assumption of 6.5% profit margin.
[b] Base year is 1976, which is the first year of DOE program expenses.

Sources: Falcone and Bjornstad (2005), DOE (2000a), Freedonia Group (2009).

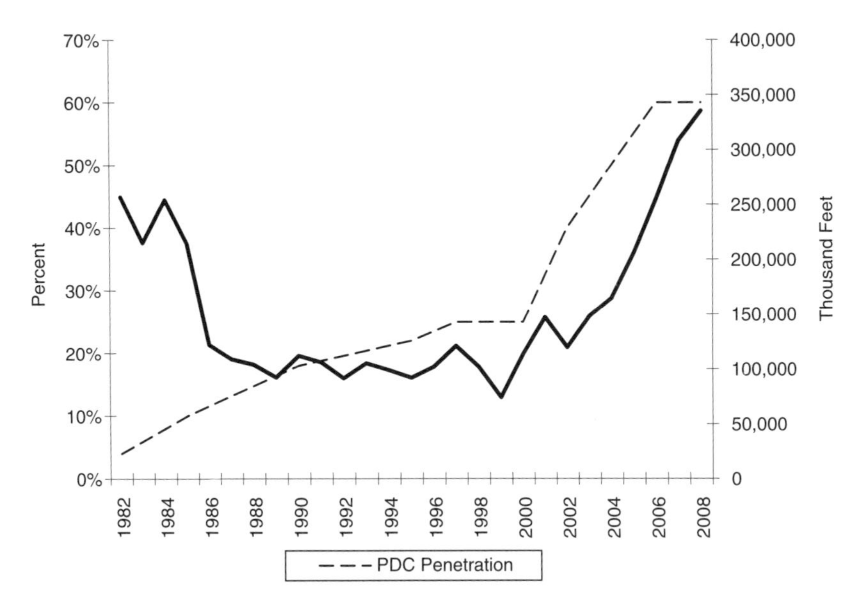

Source: Blankenship (2009).

Figure 6.2 PDC drill bits' penetration curve and total worldwide crude oil and natural gas well footage drilled with PDC bits, 1982–2008

$$\text{Cost Savings} = (\text{PDC Market Penetration})$$

$$\times (\text{Annual Oil and Gas Well Footage Drilled}) \times (\text{Savings Per Foot})$$

where

- 'PDC market penetration' is the percentage of crude oil and gas wells drilled with PDC bits.
- 'Annual well footage drilled' is the oil well footage drilled in the United States annually.
- 'Savings-per-foot' is the inflation-adjusted dollars saved per foot from drilling wells with PDC bits as compared with baseline roller cone bits.

Figure 6.2 presents the percentage of world oil and gas well footage drilled with PDC drill bits. This study assumes that drilling in the US followed a similar pattern. Market penetration started low in the early 1980s and grew greatly in the 1990s with greater use of horizontal drilling and an increase in demand for oil and gas.

Figure 6.2 also presents total developmental oil and gas well footage drilled in the US from 1982 (the year PDC drill bits were introduced to the market) to 2008. Since the oil boom of the early 1980s, there has been a significant decline in the total footage drilled; however, the total footage drilled started rising in the early 2000s and reached 350 million feet in 2008.

Table 6.10 presents the calculation of average savings per foot from drilling oil wells with PDC drill bits as compared with conventional roller-cone bits. Seven studies were located that presented savings-per-foot estimates, dating from 1982 to 1997. These studies represent 4 geographic locations and presented a total of 19 savings estimates for 7 different hole sizes for both oil and gas wells. Savings estimates were adjusted by the gross domestic product deflator and weighted by the length drilled. The average cost saving was $58.5 per foot in 2008 dollars.

Crude oil and gas well footage drilled with PDC drill bits (see Table 6.11) was calculated based on information from Figure 6.2. This footage was multiplied by cost reduction of $59 per foot to obtain annual benefits to the oil and gas extraction industry. The present value of total benefits equaled approximately $15 billion (discounted at 7 per cent).

Calculations of total benefits of PDC drill bit technology are presented in Table 6.12. Total benefits represent the sum of cost reductions to oil and gas producers and profit to PDC drill bit manufacturers. For 1982 to 2008, the introduction of PDC drill bit technology resulted in total benefits with present value of $15.1 billion (discounted at 7 per cent).

Environmental benefits

The development of PDC drill bits was one of several technological advances that in combination supported the introduction of directional drilling. Through the use of horizontal drilling, the physical footprint of drilling platforms has been reduced, resulting in less disruption of the environment. Because it would be difficult to quantify this impact, it was not included in the monetary benefits estimates.

Security benefits

The use of PDC drill bits lowered the cost of oil and gas production and potentially increased domestic supply. This may have had an effect on the US dependence on foreign oil and gas imports. However, quantifying these impacts was beyond the scope of this study.

Attribution Share

SNL played a critical role in the development and adoption of the PDC drill bit technology. Through a review of the literature and interviews

Table 6.10 Average cost reductions per foot in crude oil and natural gas wells drilled with PDC bits

Study (year)	Year sampled	area	Hole diameter (in)	Length drilled (feet)	Real savings per foot drilled (2008$)	Weighted savings per foot drilled (2008$)
Slack and Wood (1982)	1981	Texas	8 3/4	1700	20	0.7
	1981	Texas	8 3/4	731	288	4.2
	1981	Texas	8 3/4	800	309	4.9
	1981	Texas	8 3/4	3412	13	0.9
	1981	Texas	8 3/4	4590	47	4.3
	1981	Texas	8 1/2	1782	180	6.4
	1981	Louisiana	8 1/2	3860	12	0.9
	1981	Louisiana	6 3/4	4113	11	0.9
Gani (1982)	1982	Indonesia	12 1/4	3491	35	2.5
	1982	Indonesia	7 3/8	705	192	2.7
	1982	Indonesia	7 3/8	507	80	0.8
Wampler and Myhre (1990)	1990	South Texas	8 3/4	3000	9	0.6
	1990	South Texas	8 3/4	1300	6	0.2
	1990	South Texas	6 1/4	2100	7	0.3
Boudreaux and Massey (1994)	1994	Gulf of Mexico	12 1/4	2841	69	3.9
Casto (1995)	1995	Gulf of Mexico	12 1/4	2841	187	10.6
McDonald and Felderhoff (1996)	1996	Gulf of Mexico	6 1/8	2701	22	1.2
Mensa-Wilmot (1997)	1997	Gulf of Mexico	12 1/4	5814	43	5.0
	1997	Gulf of Mexico	12 1/4	3767	101	7.6
Average					86	58.5

with experts, this study found that SNL significantly contributed to the development and adoption of PDC technology by (1) developing the bit and getting it to the market, (2) overcoming many of the performance flaws and limitations, and (3) spurring the innovation that resulted in overall market success of PDC drill bits. Table 5.6 provides an overview of

Table 6.11 Benefits from using PDC drill bit technology

Year	Crude oil and natural gas exploratory and developmental well footage drilled with PDC bits[a] (thousand feet)	Savings based on well footage[b] (thousands 2008$)
1982	10 273	601 394
1983	12 897	755 006
1984	20 331	1 190 201
1985	21 402	1 252 899
1986	14 136	827 538
1987	14 380	841 823
1988	15 392	901 066
1989	15 145	886 607
1990	20 168	1 180 659
1991	19 933	1 166 902
1992	17 927	1 049 468
1993	21 442	1 255 240
1994	20 955	1 226 731
1995	20 239	1 184 815
1996	23 969	1 403 174
1997	30 265	1 771 750
1998	25 532	1 494 674
1999	18 548	1 085 822
2000	28 346	1 659 409
2001	47 839	2 800 553
2002	47 863	2 801 958
2003	66 695	3 904 406
2004	82 322	4 819 229
2005	113 184	6 625 928
2006	152 999	8 956 746
2007	184 832	10 820 288
2008	201 227	11 780 071
Undiscounted total		74 244 354
PV[c] at 7%		14 981 772

Notes:

[a] Crude Oil and Natural Gas Exploratory and Developmental Well Footage Drilled with PDC Bits = (Oil Well Footage + Gas Well Footage) x Penetration Curve (from Figure 5.2).

[b] Savings Based on Wells Footage = Crude Oil and Natural Gas Exploratory and Developmental Well Footage Drilled with PDC Bits x Average Cost Reductions per Foot in Crude Oil and Natural Gas Wells Drilled with PDC Bits ($58.5 from Table 5.2).

[c] Base year is 1976, which is the first year of DOE program expenses.

Sources: Crude Oil Developmental and Exploratory Well Footage (DOE, 2010a); Natural Gas Developmental and Exploratory Well Footage (DOE, 2010b).

Table 6.12 Total economic benefits from PDC drill bit technology (2008$)

Year	Savings based on well footage	Annual profits from sales of PDC drill bits	Total benefit
1982	$601 381	$38 264	$639 646
1983	$754 990	$46 210	$801 201
1984	$1 190 177	$70 215	$1 260 392
1985	$1 252 873	$71 739	$1 324 612
1986	$827 521	$46 355	$873 876
1987	$841 805	$45 828	$887 633
1988	$901 048	$47 425	$948 472
1989	$886 588	$44 967	$931 555
1990	$1 180 635	$57 649	$1 238 284
1991	$1 166 878	$55 027	$1 221 904
1992	$1 049 447	$48 345	$1 097 792
1993	$1 255 215	$56 574	$1 311 788
1994	$1 226 706	$54 147	$1 280 853
1995	$1 184 791	$51 233	$1 236 024
1996	$1 403 145	$59 542	$1 462 687
1997	$1 771 713	$73 870	$1 845 583
1998	$1 494 643	$61 622	$1 556 266
1999	$1 085 800	$44 117	$1 129 917
2000	$1 659 375	$65 995	$1 725 369
2001	$2 800 495	$108 925	$2 909 420
2002	$2 801 900	$107 235	$2 909 135
2003	$3 904 325	$146 291	$4 050 617
2004	$4 819 130	$175 569	$4 994 699
2005	$6 625 791	$233 611	$6 859 402
2006	$8 956 561	$305 803	$9 262 365
2007	$10 820 065	$359 166	$11 179 231
2008	$11 779 829	$382 852	$12 162 681
Undiscounted total	$74 242 828	$2 858 577	$77 101 405
PV[a] at 7%			$15 626 424

Note: [a] Base year is 1976, which is the first year of DOE program expenses.

SNL's contributions at the different stages of the PDC bit development life cycle.

SNL conducted research on bit mechanics and hydraulics that would prove useful in overcoming some of the technologies' limitations. The lab's research identified the cause of and provided a solution to catastrophic bit failure and solved the bit's spalling (or chipping off of diamond cutter) problem, resulting in solutions that became industry standards. To resolve the spalling problem, an issue caused by wear rate, SNL established an optimal bit parameter (a 20-degree back-rake angle) and designed a nozzle layout to achieve optimal cooling of the bit. Both of those innovations were adopted and widely used by bit manufacturers. SNL also created sophisticated computer code that allowed manufacturers to position the desired number of cutters on a bit (Falcone and Bjornstad, 2005).

SNL's R&D efforts, publications, and innovations brought attention and publicity to PDC drill bit technology, resulting in its overall market success. As mentioned previously, General Electric acknowledged that SNL's research helped deliver PDC drill bits to the market on time. Many officials credited the 'publicness' of Sandia's efforts as a difference maker in PDC drill bit success (Papadakis and Link, 1997). SNL's research resulted in 34 published journal articles and 42 presentations, representing the flow of information from SNL to the industry. Moreover, approximately half of the industry used one of the versions of SNL's computer code.

Furthermore, using PDCWEAR code as a basis, Amoco developed an anti-whirl drill bit, which further increased industry savings and spurred a new wave of innovations (Falcone and Bjornstad, 2005). As one expert noted, PDC drill bits were an enabling technology for the horizontal drilling[5] that is so heavily relied upon for offshore drilling (Blankenship, 2009). An estimated 60 per cent of worldwide oil and gas well footage was drilled using PDC drill bits in 2008 (Figure 6.2). A large share of this footage is attributable to horizontal drilling.

SNL also helped with initial PDC bit development and commercial introduction. General Electric acknowledges that SNL helped deliver PDC drill bits to market several years before General Electric could on its own. Thus, PDC drill bits were introduced on the eve of the drilling boom of the early 1980s, and the increased demand for drill bits during this period overcompensated for PDC bits' initial negative reputation. Had the PDC bits been introduced just a few years later, they would have failed to achieve significant market penetration because of the widespread entrenchment of roller-bit technology (Papadakis and Link, 1997; Falcone and Bjornstad, 2005).

The published literature documents that, overall, SNL played a crucial role in the innovation and market success of PDC drill bits. The PDCWEAR program code was used by almost half of the industry companies and represents a discrete piece of technology transferred from SNL to industry. Nevertheless, equally important was the incalculable amount of knowledge transferred since the beginning of SNL-industry collaboration in the 1970s, which included research on the mechanics, physics, and hydraulics of PDC bit operation. As a result, based on the observable technology transfer, findings from published papers, and interviews, this study attributes 50 per cent of the economic benefits from PDC bits to the DOE. As shown in Table 5.6, the DOE's influence varied across the stages of technology research development and commercialization, with the DOE's greatest impact occurring during system development and validation/demonstration.

This attribution estimate is consistent with previous analysis. Papadakis and Link (1997) mention that about half of industry references to major improvements during a critical stage of the PDC drill bit product cycle were to Sandia's research, and assign 50 per cent of economic benefits of PDC drill bit technology to SNL. Falcone and Bjornstad (2005) state that even though SNL's research was a critical precondition to market success, SNL partnered with industry stakeholders to bring innovation to the market. These partnerships were often equal; thus, Falcone and Bjornstad (2005) also assigned 50 per cent of economic profits to SNL.

The total benefits of PDC drill bit technology were calculated in the previous section, and, because 50 per cent of those savings are attributable to SNL, the total benefit from SNL's contributions to developing drill bit technology equaled $37.4 billion. The timeline of benefits is presented in Figure 6.3.

Benefit-Cost Analysis for PDC Drill Bits

DOE program expenditures and benefits associated with PDC drill bit research are presented in Table 6.13. These expenditures are based on a time series of appropriation items provided by DOE for EERE budgets from 1976 to 2008. The expenditures in Table 6.13 are derived from appropriation line items associated with PDC drill bit research and demonstration activities and represent approximately 1.6 per cent of EERE's total GTP budget during this time period. The present value of total expenditures (adjusted to 2008$) equaled $26.5 million discounted at 7 per cent from 1976 to 2008 ($41 million discounted at 3 per cent). DOE's investment in PDC drill bit technology yielded net benefits with a present value of $7.8 billion in net benefits ($18.4 billion discounted at 3 per cent).

Table 6.14 summarizes the results of the benefit-cost analysis. In addition to net benefits with a present value of $7.8 billion discounted at 7 per cent, the ratio of benefits relative to the DOE's expenditures was calculated (referred to as the BCR). The BCR discounted at 7 per cent is calculated to be 285.3 (437.1 discounted at 3 per cent), signifying large social benefits relative to program expenditures. The BCR decreases at higher discount rates, reflecting the timing of expenditures and benefits. IRR serves as a measure of an investment's return by comparing initial investments with

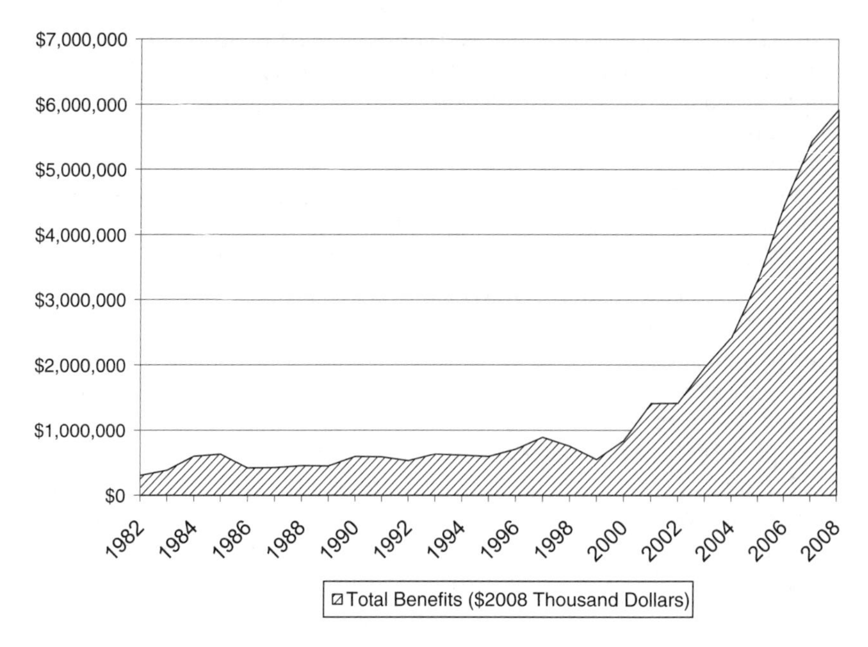

Figure 6.3 PDC drill bit technology benefits attributable to DOE, 1982–2008

discounted cash flows. The IRR for PDC drill bit technology project was 139 per cent.

CONCLUSIONS

The study used a cluster approach to generate a conservative estimate of economic performance for the GTP as a whole. The cluster analysis compared the aggregate benefits from the four selected technology areas with the investment costs of the entire program. The analysis yields net benefits with a present value of $8.1 billion (discounted at 7 per cent), a BCR of 4.9 (discounted at 7 per cent) and an IRR of 22 per cent.

The study is retrospective in that only benefits and costs through 2008 were included in the analysis. As a result, the measures of economic return calculated in this study are conservative, because in many instances, the DOE's historical R&D activities will continue to generate benefits well into the future. In addition, the nature of the cluster analysis (where total program costs are compared with benefits from a subset of selected technologies) contributes to the conservative nature of the empirical findings.

Table 6.13 PDC drill bit technology net benefits attributable to DOE, 1976–2008 (thousands 2008$)

Year	Total benefits	Program expenses	Net benefits
1976	$0	$2 081	−$2 081
1977	$0	$2 081	−$2 081
1978	$0	$2 081	−$2 081
1979	$0	$2 081	−$2 081
1980	$0	$2 081	−$2 081
1981	$0	$2 081	−$2 081
1982	$319 823	$2 081	$317 742
1983	$400 600	$2 081	$398 519
1984	$630 196	$2 081	$628 115
1985	$662 306	$2 081	$660 225
1986	$436 938	$2 081	$434 857
1987	$443 817	$1 844	$441 973
1988	$474 236	$1 683	$472 553
1989	$465 778	$1 601	$464 177
1990	$619 142	$1 490	$617 652
1991	$610 952	$1 586	$609 366
1992	$548 896	$1 575	$547 321
1993	$655 894	$1 455	$654 439
1994	$640 427	$1 335	$639 092
1995	$618 012	$1 341	$616 671
1996	$731 344	$1 474	$729 870
1997	$922 792	$1 558	$921 234
1998	$778 133	$1 817	$776 316
1999	$564 958	$1 979	$562 979
2000	$862 685	$2 045	$860 640
2001	$1 454 710	$2 188	$1 452 522
2002	$1 454 567	$2 188	$1 452 379
2003	$2 025 308	$2 188	$2 023 120
2004	$2 497 349	$2 188	$2 495 161
2005	$3 429 701	$2 188	$3 427 513
2006	$4 631 182	$2 188	$4 628 994
2007	$5 589 616	$2 188	$5 587 428
2008	$6 081 340	$2 188	$6 079 152
Undiscounted totals	$38 550 702	$63 178	$38 487 524
PV[a] at 7%	$7 813 212	$26 461	$7 786 751
PV[a] at 3%	$18 514 201	$41 015	$18 473 186

Note: [a] Base year is 1976, which is the first year of DOE program expenses.

Table 6.14 PDC drill bit technology benefit-cost analysis results

	Net benefits (thousands $2008)	Benefit-cost ratio	Internal rate of return
PV[a] at 7%	7786751	295.3	139%
PV[a] at 3%	18473186	451.4	

Note: [a] Base year is 1976, which is the first year of DOE program expenses.

It should also be noted that additional benefits are likely to be associated with the four technologies that could not be quantified given this study's timing and resources. For example, environmental health benefits only capture the impact of reducing emissions of particulate matter, nitrous oxides, and sulfur oxides, and do not reflect other environmental health benefits associated with reductions of other pollutants (such as CO_2, mercury).

NOTES

1. Liquid-dominated resources are those in which liquid has not vaporized into steam (as opposed to vapor-dominated resources).
2. This is true in instances where PDC bits can be applied. PDC bits are less successful in high-temperature hard and fractured rock formations. Because the principal wear mechanism of PDC drill bits is frictional heating, high temperatures only aggravate wear.
3. Here and throughout the chapter, unless otherwise noted, PVs are discounted at 7 per cent.
4. The increased productivity associated with binary cycle over flash at this temperature range is achieved at negligible incremental cost. Thus, the economic benefit is the value of the power generated. For the lower temperature range, because the cost of binary power generation is comparable to the cost of coal power generation there is no economic benefit, but there are environmental benefits.
5. Horizontal drilling is the process of drilling and completing, for production, a well that begins as a vertical or inclined linear bore that extends from the surface to a subsurface location just above the target oil or gas reservoir, called the 'kickoff point,' then bears off on an arc to intersect the reservoir at the 'entry point,' and thereafter, continues at a near-horizontal attitude tangent to the arc to substantially or entirely remain within the reservoir until the desired bottom hole location is reached (DOE, 1993). The total linear footage used in the benefits calculations captures both vertical and linear drilling.

7. Investments in vehicle combustion engine technologies

INTRODUCTION

DOE has had a history of involvement in vehicle technologies, including being one of the major participants involved in President Clinton's establishment of the Partnership for a New Generation of Vehicles (PNGV) in 1993. Joining the DOE in this partnership were seven other government agencies (the Departments of Commerce, Defense, Interior, and Transportation; the NSF; NASA; and the EPA); national laboratories; and the Chrysler Corporation, the Ford Motor Company, and General Motors (through the United States Council for Automotive Research). The goals of the PNGV were as follows:

> (1) to improve national manufacturing competitiveness, (2) to implement commercially viable technologies that increase the fuel efficiency and reduce the emissions from conventional vehicles, and (3) to develop technologies for a new class of vehicles with up to three times the fuel efficiency of 1994 midsize family sedans (80 mpg) while meeting emission standards and without sacrificing performance, affordability, utility, safety, or comfort. (NRC, 2001, p. 146)

The Bush Administration modified the PNGV program in 2001 and adopted a new focus through the creation of FreedomCAR in 2003. One emphasis of FreedomCAR was on hydrogen fuel cells (PNGV, 2009).

The counterpart to the PNGV for passenger vehicles was the 21st Century Truck Partnership, announced in April 2000. The goals of this government program were to improve fuel efficiency in long-haul trucks, increasing Class 7 and Class 8 highway truck fuel efficiency by 20 per cent, from the current 42 per cent thermal efficiency to 50 per cent thermal efficiency by 2010 and 55 per cent thermal efficiency by 2013 and to lower emission beyond the expected standard for 2010 (NRC, 2008). Initially, the 21st Century Truck Partnership was under the administrative authority of the US Army Tank-Automotive Research and Development Command within the Department of Defense. But, in November 2002, authority over the Partnership passed to the DOE, specifically to the VTP under EERE: 'DOE [was] assigned to lead the federal R&D component of this program

Table 7.1 Activity areas within the VTP

Subprogram	Description
Hybrid and Vehicle Systems Technologies	Analysis and testing activities that provide support and guidance for many cutting-edge automotive and truck technologies now under development
Energy Storage Technologies	Critical enabling battery technologies for the development of advanced, fuel-efficient light- and heavy-duty vehicles
Power Electronics and Electrical Machines Technologies	Motors, inverters/converters, sensors, control systems, and other interface elements that are critical to hybrid electric and fuel cell vehicles
Advanced Combustion Engine R&D (ACE R&D)	Technologies that contribute to more efficient, advanced internal combustion engines in light-, medium-, and heavy-duty vehicles
Fuels and Lubricants Technologies	Fuel and lubricant options that are cost-competitive, enable high fuel economy, deliver lower emissions, and contribute to petroleum displacement
Materials Technologies	Lightweight, high-performance materials that can play an important role in improving the efficiency of transportation engines and vehicles
EPAct	Programs in support of the Energy Policy Act of 1992 (EPAct), which was passed to reduce The US's reliance on foreign petroleum and improved air quality
Educational Activities	Collegiate programs that help encourage engineering and science students to pursue careers in the transportation sector

Source: EERE (2009b).

because of the close alignment of the stated 21st Century Truck Program goals and research objectives with DOE's mission "to foster a secure and reliable energy system that is environmentally and economically sustainable"' (NRC, 2008, p. 9).

As previously mentioned, EERE was formed in 2001 when the Office of Conservation and Solar Energy was renamed and reorganized. EERE includes ten energy programs. The Vehicle Technologies Program (VTP) encompasses eight broad subprogram areas, as listed in Table 7.1. In the most general terms, the ACE R&D subprogram sponsors R&D to address technical barriers to the commercialization of higher-efficiency internal combustion engines used in passenger and commercial vehicles (see DOE [2003]).

The research areas in the ACE R&D subprogram are Laser Diagnostics and Optical Engine Technologies, Combustion Modeling, Combustion and Emission Control, and Solid State Energy Conversion. This case study relates to the application of laser and optical diagnostics and combustion modeling to heavy-duty diesel engines.

Laser and optical diagnostics and combustion modeling are two research areas for which there are measurable milestones and outcomes (e.g., improvements in brake thermal efficiency and miles per gallon) that are directly associated with the ACE R&D subprogram's research.

The emphasis on heavy-duty diesel trucks reflects the fact that trucking is a vital industry to the US economy and to national income. Trucks account for about 25 per cent of the transportation industry's total revenues. Based on the Economic Census of 2002: '[T]he truck transportation industry consisted [in 2002] of more than 112,698 separate establishments, with total revenues of $165 billion. These establishments employ 1,437,259 workers, who take home an annual payroll of $47 billion' (NRC, 2008, p. 9).

Trucks also account for nearly 58 per cent of total highway transportation energy consumption; heavy-duty trucks account for nearly 24 per cent of total highway transportation energy consumption (Davis et al., 2009). The number of heavy-duty diesel truck registrations has increased since 1970. In 1970, 905,000 heavy-duty diesel trucks were registered and they were driven 35.1 billion miles; in 2007, the number of registered heavy-duty diesel trucks rose to 2.2 million and they were driven 145.0 billion miles.[1]

OVERVIEW OF THE BENEFIT-COST ANALYSIS

This case study summarizes a retrospective benefit-cost evaluation analysis; only benefits and costs through 2007 are considered, although laser and optical diagnostics and combustion modeling will affect diesel fuel engine efficiency into the future. As a result of this retrospective focus, and other assumptions discussed below, the findings presented herein are conservative.

The discussion that follows identifies, documents, and validates three categories of benefits: economic benefits, environmental and health benefits, and energy security benefits. Economic benefits for fuel reduction are quantified in monetary terms, as are the health benefits. Environmental emission reduction benefits are quantified but not monetized. The security benefits are also described using quantitative, nonmonetary measures.

Categories of benefits quantified in monetary terms, which are associated with research in and the application of laser and optical diagnostics and combustion modeling applied to heavy-duty diesel engines, are compared with the total research costs of the entire ACE R&D subprogram's research areas, including the research costs associated with the

Combustion Research Facility (CRF). These comparisons are then calculated using traditional economic evaluation metrics.

BUDGET HISTORY OF ADVANCED COMBUSTION ENGINEERING R&D

Table 7.2 shows the aggregate annual appropriation budgets for both the VTP and the ACE R&D subprogram. It also shows the annual DOE Office of Science budgets for the cross cutting research programs that are related to combustion and that are within the CRF budget. Data are missing for several years. Approximations for these missing data are shown in italics, with explanations about the approximations in the notes following the table. For reference purposes only, the VTP budget is reported in Column (2). In 2008, the ACE R&D subprogram budget was nearly 21 per cent of the VTP budget.

The sum of the cost data for the ACE R&D subprogram and for the CRF, by year, is the appropriate cost basis for research that led to advances in laser and optical diagnostics and combustion modeling related to heavy-duty diesel engines. There, all data are in inflation-adjusted (real) 2008 dollars (see Columns 7 and 8 of Table 7.2). The conversion of actual (nominal) costs to real costs is through the gross domestic product (GDP) implicit price deflator, shown in Column (5) with 2005 as the base year and in Column (6) with 2008 as the base year.

The sum of the ACE R&D subprogram and CRF budgets overstates the costs of laser and optical diagnostics and combustion modeling research for several reasons. First, laser and optical diagnostics and combustion modeling are only two of the four research areas within the ACE R&D subprogram. In 1986 only about 33 per cent of the ACE R&D's engine budget was focused on diesel engines for light- and heavy-duty applications, but that percentage increased over time to about 80 per cent by the mid 1990s and was even larger by 2007; with heavy-duty applications of interest in this study about a third of the total.

Table 7.3 summarizes the specific costs that were compared with the benefits attributable to laser and optical diagnostics and combustion modeling. Specifically, the total of the ACE R&D subprogram research costs and the CRF research costs (hereafter, simply ACE R&D subprogram costs) are used in the benefit-cost evaluation calculations that follow. These costs began in 1986, the first year of available cost data for the ACE R&D subprogram and the CRF; more importantly, this is also the approximate date when laser and optical diagnostics and combustion modeling research started to be applied to heavy-duty diesel engines.

Table 7.2 Annual appropriations for the VTP, the ACE R&D subprogram, and the CRF ($ millions)

(1) Year	(2) VTP budget	(3) ACE R&D sub-program budget	(4) CRF budget[a]	(5) GDP implicit price deflator (2005=100)	(6) GDP implicit price deflator (2008=100)	(7) Inflation-adjusted ACE R&D sub-program budget (2008$)	(8) Inflation-adjusted CRF budget (2008$)
1976	$12540			35489	32714		
1977	$28425			37751	34799		
1978	$63798*			40400	37241		
1979	$99170			43761	40339		
1980	$110500			47751	44017		
1981	$105050			52225	48141		
1982	$58944			55412	51079		
1983	$53856			57603	53099		
1984	$64900			59766	55093		
1985	$61772			61576	56761		
1986	$57457	*$15897***	$3250	62937	58016	27402	5602
1987	$55393	*$17316***	$3540	64764	59700	29005	5930
1988	$51360	*$17157***	*$3508**	66988	61750	27785	5680
1989	$54330	*$16998***	$3475	69518	64082	26525	5423
1990	$68394	$17257	$3719	72201	66555	25929	5588
1991	$83564	$15760	$4300	74760	68914	22869	6240
1992	$109282	$16657	$4390	76533	70548	23611	6223
1993	$138632	$14818	$4379	78224	72107	20550	6073
1994	$177249	$12949	$4171	79872	73626	17587	5665
1995	$191065	$10440	$4171	81536	75160	13890	5549
1996	$174288	$16524	$4714*	83088	76591	21574	6154
1997	$172457	$19263	$5256	84555	77943	24714	6743
1998	$189972	$18318	$5161	85511	78824	23239	6547
1999	$198665	$36976	$5024	86768	79983	46230	6281
2000	$228756	$46750	$4736	88647	81715	57211	5796
2001	$251462	$52205	$5463	90650	83561	62475	6538
2002	$181352	$47160	$5377	92118	84915	55538	6332
2003	$174171	$55267	$5935	94100	86742	63714	6842
2004	$172395	$52736	$5892	96770	89203	59119	6605
2005	$161326	$48480	$6437	100	92180	52593	6983
2006	$178351	$40594	$6251	103257	95183	42649	6567
2007	$183580	$48346	$7648	106214	97908	49379	7811
2008	$208359	$43443	$6755	108483	100	43443	6755

Table 7.2 (continued)

Notes: [a] Two years of CRF construction began in 1978, with early years of operation beginning in 1980 through 1985. The complete funding information for those years is unknown.
When data are not available for a particular year/program, the cell is blank.
* denotes values that were constructed as the average for the juxtaposed years.
** denotes values that were constructed on the basis of the average ratio of the CRF budget to the ACE R&D subprogram budget for all available years.
Column (4) represents DOE Office of Science funding for cross-cutting research programs that are related to combustion and that are within the CRF.
Column (6) = Column (5) / (108.483 / 100).
Column (7) = Column (3) / (Column (6) / 100).
Column (8) = Column (4) / (Column (6) / 100).
Year 2008 shown to benchmark the GDP deflator in Column (6).

Sources: Nominal budget data in Columns (2)–(4) provided by EERE.
GDP Implicit Price Deflator (2005=100) from DoC (2009).

ESTIMATION OF BENEFITS

Economic Benefits

Economic benefits were quantified in terms of improved fuel efficiency of heavy-duty diesel trucks associated with laser and optical diagnostics and combustion modeling technologies.

One widely used measure of fuel efficiency is brake thermal efficiency (BTE), defined as the percentage of fuel energy converted into work energy during the energy cycle.[2] Simply, the more efficient the combustion process, the greater the amount of fuel energy that is converted into energy, as opposed to being emitted as exhaust, and hence the greater the miles per gallon.

Figure 7.1 shows BTE for heavy-duty diesel trucks from 1960 to the present. Of particular importance in the figure are the increase in BTE from 1960 (the first year of available data) to 2002, along with the more rapid increase in BTE during the mid 1980s and 1990s, and the precipitous drop in BTE in 2002. The data that underlie Figure 7.1 are found in Table 7.4.

The trend in BTE in Figure 7.1 is important in the calculation of economic benefits below. In particular, changes in EPA regulations have influenced BTE over time. In the statistical analysis related to the calculation of economic benefits, that influence is held constant or controlled in the relationship between BTE and related fuel savings.

EPA applied its first emissions standards to heavy-duty diesel engines beginning with model year 1974, setting upper limits for NO_x, sulfur

Table 7.3 Cost data used in the evaluation of economic benefits ($ millions)

(1) Year	(2) ACE R&D subprogram (2008$)	(3) CRF (2008$)	(4) Total (2008$)
1986	27402	5602	33004
1987	29005	5930	34935
1988	27785	5680	33465
1989	26525	5423	31948
1990	25929	5588	31517
1991	22869	6240	29109
1992	23611	6223	29834
1993	20550	6073	26623
1994	17587	5665	23252
1995	13890	5549	19439
1996	21574	6154	27728
1997	24714	6743	31457
1998	23239	6547	29786
1999	46230	6281	52511
2000	57211	5796	63007
2001	62475	6538	69013
2002	55538	6332	61870
2003	63714	6842	70556
2004	59119	6605	65724
2005	52593	6983	59576
2006	42649	6567	49216
2007	49379	7811	57190

Note: Column (4) = Column (2) + Column (3).

dioxide (SO_x) and hydrocarbon emissions. Standards were first placed on PM for vehicles in model year 1988. In 2000, EPA finalized the rules that require additional reductions in NO_x for newly manufactured highway diesel engines. The higher standard was intended to apply to heavy-duty diesel trucks beginning in model year 2004. In 1998, EPA, the US Department of Justice, and the California Air Resources Board settled a lawsuit over several manufacturers' installation of defeat devices in heavy-duty diesel trucks. These engine performance-enhancing devices also increased NO_x emissions during specific performance intervals, thereby evading proper emissions testing and violating the emissions standards passed in 1974. As part of the settlement (referred to as the consent decree),

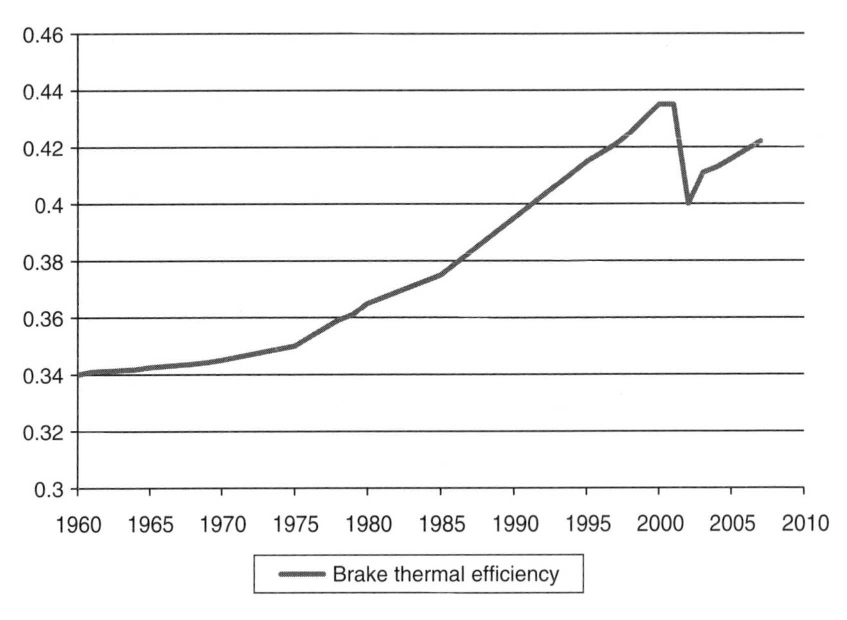

Figure 7.1 Trend in BTE over time

these manufacturers agreed to meet the aforementioned 2004 standards 15 months early (in October 2002) and to abandon the use of injection timing devices. As a result of this accelerated schedule, there was insufficient time for manufacturers to develop technologies that could both preserve engine efficiency and reduce emissions (Siebers, 13 October 2009; scientists at Cummins and Detroit Diesel). Fuel efficiency, therefore, decreased during this adjustment period, as research resources were shifted away from improved fuel efficiency to meet these emissions standards (Singh, 2000; Kalish, 2009).

Yet another new standard for NO_x and SO_x was implemented for heavy-duty diesel trucks for model year 2007, cutting emissions by about 90 per cent from the originally scheduled model year 2004 threshold. Tighter emissions limits for PM were imposed in the same year. The regulations were imposed under the authority of the Clean Air Act and included a requirement to reduce sulfur content of diesel fuel by 97 per cent (CRS, 2001; CATF, 2005). Table 7.5 summarizes these periods of EPA regulations.

As previously discussed, increases in BTE are positively related to increases in miles per gallon and decreases in fuel consumption, all else remaining constant (Siebers, 8 September 2009).[3] Thus, the time series of BTE in Figure 7.1 and Table 7.4 will be a portion of the primary data used in the calculation of economics benefits.

Table 7.4 Values of BTE shown in figure 7.1

(1) Year	(2) BTE	(1) Year (cont.)	(2) BTE (cont.)	(1) Year (cont.)	(2) BTE (cont.)
1960	0.340	1976	0.353	1992	0.403
1961	0.341	1977	0.356	1993	0.407
1962	0.341	1978	0.359	1994	0.411
1963	0.342	1979	0.361	1995	0.415
1964	0.342	1980	0.365	1996	0.418
1965	0.343	1981	0.367	1997	0.421
1966	0.343	1982	0.369	1998	0.425
1967	0.343	1983	0.371	1999	0.430
1968	0.343	1984	0.373	2000	0.435
1969	0.344	1985	0.375	2001	*0.435*
1970	0.345	1986	0.379	2002	0.400
1971	0.346	1987	0.383	2003	*0.411*
1972	0.347	1988	0.387	2004	*0.413*
1973	0.348	1989	0.391	2005	*0.416*
1974	0.349	1990	0.395	2006	*0.419*
1975	0.350	1991	0.399	2007	*0.422*

Sources: Aneja et al. (2009) and Kalish (2009). These sources report BTE values only for selected years because of the confidential nature of such data. Intervening years of data were interpolated and estimated (shown in italics) on the basis of extensive discussions with industry scientists from Cummins Engine and Detroit Diesel Corporation.

Table 7.5 Summary of EPA NO_x, PM and SO_x emission regulations on heavy-duty diesel trucks

(1) Year	(2) NO_x (g/hp-hr)	(3) PM (g/hp-hr)	(4) SO_x (ppm)
1994	5.0	0.25	500
1998	4.0	0.10	500
2002	2.5	0.10	500
2007	1.2	0.01	500
2010	0.2	0.01	15

Note: NO_x and PM are measured as emissions per unit power demand. Emissions are measured in grams (g) and power demand is measured as horsepower (hp) per hour (hr). SO_x is measured as parts per million (ppm).

Sources: Aneja et al. (2009) and CATF (2005).

Based on data provided during extensive telephone interviews with a scientist at each of three companies, Caterpillar, Cummins Engine, and Detroit Diesel Corporation,[4] which collectively account for about 75 per cent of the heavy-duty diesel engines manufactured in the US,[5] the consensus opinion from these experts is that:

- The impact of the ACE R&D subprogram's research on and application of laser and optical diagnostics and combustion modeling, which began in 1986, had a measurable impact on the BTE of heavy-duty diesel engines not later than 1995.[6,7,8]
- Without the ACE R&D subprogram's research in and application of laser and optical diagnostics and combustion modeling to heavy-duty diesel engines, BTE from 1995 through 2007 would have been 4.5 per cent lower per year than shown in Figure 3.2.[9,10] This percentage is a critical datum in the calculation of economic benefits. Figure 3.3 illustrates this assumption.[11]
- Without the ACE R&D subprogram's research in and application of laser and optical diagnostics and combustion modeling, the US diesel engine industry would not have been able to conduct the research necessary to duplicate the application of these technologies to heavy-duty diesel engines, even with the research assistance of universities. The industry could not have absorbed the requisite capital cost of R&D and could not have achieved the needed economies of scale in R&D to warrant the effort.[12]

Without the ACE R&D subprogram's research in and application of laser and optical diagnostics and combustion modeling, the next-best technology would be the state-of-the-art in diesel engine design and related BTE that existed prior to 1995. Thus, the next-best alternative technology is represented in Figure 7.2 by the line from 1995 through 2007 labeled 'BTE Absent ACE R&D Research on Diagnostics and Modeling.'

The data underlying the counterfactual situation of an absence of the ACE R&D subprogram's research in and application of laser and optical diagnostics and combustion modeling came from interview information with scientists at Caterpillar, Cummins Engine, and Detroit Diesel Corporation.

The measured increase in BTE of heavy-duty diesel engines from 1995 through 2007 is 100 per cent attributable to the ACE R&D subprogram's research in and application of laser and optical diagnostics and combustion modeling. That is, in the absence of the ACE R&D subprogram's research, the US diesel engine industry would not have been able to conduct the research necessary to duplicate the resultant technologies, even with the research assistance of universities.

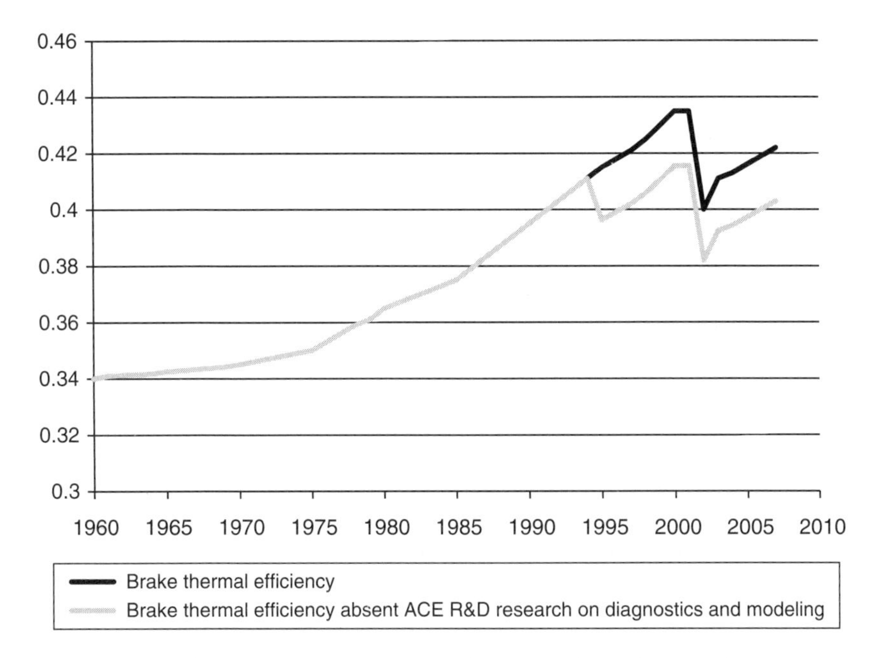

Figure 7.2 Trend in BTE over time with counterfactual scenario

This assumption of 100 per cent attribution follows from three independent sources: a documented argument used to justify the initial creation of the CRF, insight from DOE scientists and industry experts about industry's research capabilities at that time, and economic theory.

Regarding the argument used to justify the initial creation of the CRF, McLean (22 October 2009) is of the opinion that the level of knowledge and complexity required to develop and implement laser and optical diagnostics and combustion modeling measurements was too costly for industry to have developed on its own. Even under the hypothetical assumption that industry could have funded such an endeavor, including the cost of the equipment, industry would still not have been able to achieve the synergies among scientists from various fields that were achieved at the DOE-funded CRF. According to McLean (22 October 2009), the capital costs to replicate the knowledge from the ACE R&D subprogram's research in laser and optical diagnostics and combustion modeling were prohibitive from industry's perspective. And, as discussed with respect to the history of the CRF, the initial rationale for the facility was that companies could not incur the capital costs to learn detailed information about the combustion process (Carlisle et al., 2002).

Of course, the engine companies have contributed to improvements in BTE over time, but the measured increase in BTE above the counterfactual level posited herein is 100 per cent attributable to the ACE R&D subprogram's research.

Regarding the insights about industry capabilities from DOE scientists and industry experts, the discussion above explains that the next-best technology would be the state-of-the-art in diesel engine design and related BTE that existed prior to 1995. The consensus opinion of scientists and industry experts interviewed was that BTE of heavy-duty truck engines from 1995 through 2007 was 4.5 per cent lower than it would have been, without the DOE's involvement.

Finally, the assumption of 100 per cent attribution follows from the theoretical discussion in Chapter 2 regarding the economic rationale for government support of combustion research. One factor that creates barriers to innovation that lead to technological market failure is high capital costs to undertake the underlying R&D.

A statistical approach was adopted for the calculation of the economic benefits associated with the ACE R&D subprogram's research in and application of laser and optical diagnostics and combustion modeling. This statistical approach depends on the opinion of industry experts (discussed above) that in the absence of the ACE R&D subprogram's diagnostics and modeling research, BTE from 1995 to the present would be 4.5 per cent lower than actual BTE over that same period, as illustrated in Figure 7.2.

To calculate the statistical relationship between changes in BTE in year t (BTE_t) and changes in the fuel economy of heavy-duty diesel engines in year t measured by miles per gallon in year t (MPG_t), holding constant the influence of the EPA regulations in Table 7.5 on fuel economy, the following regression model was estimated:

$$MPG_t = \alpha + \beta_1 BTE_t + \beta_2 d_{94} + \beta_3 d_{98} + \beta_4 d_{02} + \varepsilon_t \quad (7.1)$$

where d_{94}, d_{98}, and d_{02} are binary variables equal to 1 for EPA regulation periods (see Table 7.5), and where ε_t is a normal and randomly distributed error term.

Fundamental to quantifying the economic benefits attributable to the ACE R&D subprogram's research in and application of laser and optical diagnostics and combustion modeling is the calculation of reduced fuel consumption per year from 1995 through 2007 on heavy-duty diesel engines. Table 7.6 shows diesel fuel consumption and fuel economy (miles per gallon [MPG]) over time. The data for MPG_t in equation (7.1) are in Column (5). These available data related to fuel economy, or MPG, are specific to all vintages of Class 7 and Class 8 heavy-duty diesel trucks

Table 7.6 Trends and summary statistics on heavy-duty diesel trucks

(1) Year	(2) Registrations (thousands)	(3) Vehicle travel (million miles)	(4) Fuel consumption (million gallons)	(5) Fuel economy (MPG)
1970	905	35 134	7 348	4.781
1971	919	37 217	7 595	4.900
1972	961	40 706	8 120	5.013
1973	1 029	45 649	9 026	5.058
1974	1 085	45 966	9 080	5.062
1975	1 131	46 724	9 177	5.091
1976	1 225	49 680	9 703	5.120
1977	1 240	55 682	10 814	5.149
1978	1 342	62 992	12 165	5.178
1979	1 386	66 992	12 864	5.208
1980	1 417	68 678	13 037	5.268
1981	1 261	69 134	13 509	5.118
1982	1 265	70 765	13 583	5.210
1983	1 304	73 586	13 796	5.334
1984	1 340	77 377	14 188	5.454
1985	1 403	78 063	14 005	5.574
1986	1 408	81 038	14 475	5.598
1987	1 530	85 495	14 990	5.703
1988	1 667	88 551	15 224	5.817
1989	1 707	91 879	15 733	5.840
1990	1 709	94 341	16 133	5.848
1991	1 691	96 645	16 809	5.750
1992	1 675	99 510	17 216	5.780
1993	1 680	103 116	17 748	5.810
1994	1 681	108 932	18 653	5.840
1995	1 696	115 451	19 777	5.838
1996	1 747	118 899	20 192	5.888
1997	1 790	124 584	20 302	6.137
1998	1 831	128 159	21 100	6.074
1999	2 029	132 384	24 537	5.395
2000	2 097	135 020	25 666	5.261
2001	2 154	136 584	25 512	5.354
2002	2 277	138 737	26 480	5.239
2003	1 908	140 160	23 815	5.885
2004	2 010	142 370	24 191	5.885
2005	2 087	144 028	27 689	5.202
2006	2 170	142 169	28 107	5.058
2007	2 221	145 008	28 515	5.085

Note: Fuel economy was reported to one significant digit. It was recalculated to three significant digits based on reported data for fuel consumption and miles of vehicle travel for this table and for use in the economic evaluation analysis.

Table 7.7 Regression results from the estimation of equation (7.1)

(1) Variable	(2) Estimated coefficient	(3) Standard error	(4) t-value
BTE	0.153*	0.018	8.49
d_{94}	−0.218	0.142	−1.54
d_{98}	−0.564*	0.146	−3.86
d_{02}	−0.661*	0.113	−5.86
constant	−0.283	0.666	−0.42
R^2	0.82		
D-W	1.91		
n	38		

Notes:
* = significant at the 0.01 level or greater.
Autocorrelation corrected using the Yule–Walker estimation method.

Source: Davis et al. (2009) as recommended by the ACE R&D staff.

registered, by year, not to only new ones. Because changes in BTE in year t will have an impact only on new heavy-duty diesel engines in trucks registered in year t, the estimated coefficient on BTE_t in equation (7.1) implicitly controls for the fact that available data on MPG are not vintage/class specific.

The regression results from the estimation of equation (7.1) using data from 1970 through 2007 are reported in Table 7.7. The estimated coefficient on BTE_t is positive, as expected, and statistically significant at the 0.001 level or better. Also, the estimated coefficients on each of the binary variables, denoting various periods of EPA regulation, are negative, as expected. The estimated coefficient on d_{94} is not statistically significant at a conventional level (it is statistically significant at the 0.15 level), but the estimated coefficients on d_{98} and d_{02} are statistically significant at the 0.001 level or better.

These regression results show a positive correlation between BTE and MPG, holding constant periods of EPA regulation. The regression R^2 is 0.82, implying that 82 per cent of the variation in MPG over time is explained by variation in BTE over time, or only 18 per cent of the variation in MPG has not been explained by the model in equation (7.1).[13] The estimated coefficient on BTE is 0.153. Based on this estimated coefficient, a one-unit change in BTE is associated with a 0.153 unit change in MPG, holding constant periods of regulation. Thus:

$$\partial MPG / \partial BTE = 0.153. \qquad (7.2)$$

Industry experts were of the opinion that in the absence of EERE's investments in the ACE R&D subprogram's research in laser and optical diagnostics and combustion modeling, BTE of heavy-duty truck engines from 1995 through 2007 would have been 4.5 per cent lower than it actually was, as shown in Figure 7.2. Thus, BTE without the ACE R&D subprogram's research in laser and optical diagnostics and combustion modeling equals BTE times (1 – 0.045) from 1995 through 2007.[14] The decrease in BTE resulting from the counterfactual absence of these technologies is, therefore, the difference between actual or observed BTE and BTE without the ACE R&D subprogram's research in laser and optical diagnostics and combustion modeling.

Based on the regression results in Table 7.7 (as interpreted in equation [7.2]), the decrease in MPG that would have occurred under the counterfactual absence of the ACE R&D subprogram's research in laser and optical diagnostics and combustion modeling equals the decrease in actual BTE times 0.153.[15] The calculations for these counterfactual-decreased MPG values are shown in Table 7.8. All other relevant data for the following calculation of these fuel efficiency benefits are shown in Table 7.9.

The calculated data in Column (5) of Table 7.8 are critical to the calculated economic benefits attributable to the ACE R&D subprogram's research in laser and optical diagnostics and combustion modeling. The calculations follow directly from the bulleted assumptions stated above, especially the assumptions related to the next-best alternative technology and attribution.

In Table 7.9, actual fuel consumption (Column [3]), and actual fuel economy (Column [4]) reflect fuel efficiency from heavy-duty diesel engines that embody the ACE R&D subprogram's research in laser and optical diagnostics and combustion modeling. Absent these technologies from 1995 through 2007, actual fuel economy (Column [4]) would be lower by the amount shown in Column (5). Lower fuel economy, or the miles per gallon that would have existed under the counterfactual situation of no laser and optical diagnostics and combustion modeling technologies, is in Column (6). Fuel consumption under the counterfactual situation (and under the implicit assumption that vehicle miles are independent of DOE research) is in Column (7). Therefore, reduced fuel consumption that can be fully attributable to the ACE R&D subprogram's research in laser and optical diagnostics and combustion modeling in heavy-duty diesel engines is in Column (8). Over the years 1995 through 2007, a total of 17,552 million gallons of diesel fuel have been saved.

Table 7.10 summarizes the monetary value of the economic benefits associated with reduced fuel consumption, by year. Fuel savings in the

Table 7.8 Calculation of decrease in fuel economy absent the ACE R&D subprogram's technologies

(1) Year	(2) BTE	(3) BTE absent ACE R&D subprogram technologies	(4) Decrease in BTE absent ACE R&D subprogram technologies	(5) Decrease in fuel economy absent ACE R&D subprogram technologies (MPG)
1995	41.5	39.633	1.867	0.286
1996	41.8	39.919	1.881	0.288
1997	42.1	40.206	1.894	0.290
1998	42.5	40.588	1.912	0.293
1999	43.0	41.065	1.935	0.296
2000	43.5	41.543	1.957	0.299
2001	43.5	41.543	1.957	0.299
2002	40.0	38.200	1.800	0.275
2003	41.1	39.251	1.849	0.283
2004	41.3	39.442	1.858	0.284
2005	41.6	39.728	1.872	0.286
2006	41.9	40.015	1.885	0.288
2007	42.2	40.301	1.899	0.291

Notes:
Column (3) = Column (2) x (1 – 0.045).
Column (4) = Column (2) – Column (3).
Column (5) = Column (4) x 0.153.

table are valued in terms of the average annual market price of diesel fuel. From 1995 through 2007, these savings totaled $34,496 million (2008$).

Environmental and Health Benefits

The environmental benefits associated with the 17.6 billion gallons of diesel fuel saved between 1995 through 2007 are quantified in terms of reduced CO_2 emissions, but these greenhouse gas emission reductions are not monetized. From 1995 through 2007, the emission of CO_2 from heavy-duty diesel trucks was reduced by 177.3 million metric tons, and this reduction is fully attributable to the DOE's investment in the research in laser and optical diagnostics and combustion modeling.

The environmental benefits associated with the 17.6 billion gallons of diesel fuel saved are further quantified in terms of reduced emissions of NO_x, PM, and SO_x using EPA's COBRA model, as discussed in the Appendix to this book. Those reductions are shown in Table 7.11.

Table 7.9 Reduced fuel consumption with the ACE R&D subprogram's laser and optical diagnostics and combustion modeling technologies (rounded)

(1) Year	(2) Vehicle travel (million miles)	(3) Fuel consumption (million gallons)	(4) Fuel economy (MPG)	(5) Decrease in fuel economy absent ACE R&D subprogram's technologies (MPG)	(6) Fuel economy absent ACE R&D subprogram's technologies (MPG)	(7) Fuel consumption absent ACE R&D subprogram's technologies (million gallons)	(8) Reduced fuel consumption with ACE R&D subprogram's technologies (million gallons)
1995	115451	19777	5.838	0.286	5.552	20794	1017
1996	118899	20192	5.888	0.288	5.600	21232	1040
1997	124584	20302	6.137	0.290	5.847	21307	1005
1998	128159	21100	6.074	0.293	5.781	22169	1069
1999	132384	24537	5.395	0.296	5.099	25963	1426
2000	135020	25666	5.261	0.299	4.962	27211	1545
2001	136584	25512	5.354	0.299	5.055	27020	1508
2002	138737	26480	5.239	0.275	4.964	27949	1469
2003	140160	23815	5.885	0.283	5.602	25020	1205
2004	142370	24191	5.885	0.284	5.601	25419	1228
2005	144028	27689	5.202	0.286	4.916	29298	1609
2006	142169	28107	5.058	0.288	4.770	29805	1698
2007	145008	28515	5.085	0.291	4.794	30248	1733
Total							17552

Notes:
Column (6) = Column (4) – Column (5).
Column (7) = Column (2) / Column (6).
Column (8) = Column (7) – Column (3).

Table 7.10 *Economic benefits of reduced fuel consumption from the ACE R&D subprogram's research on laser and optical diagnostics and combustion modeling technologies (rounded)*

(1) Year	(2) Reduced fuel consumption with ACE R&D subprogram's technologies (million gallons)	(3) Average retail price diesel fuel ($ per gallon)	(4) Dollar value of reduced fuel consumption (millions $)	(5) GDP implicit price deflator (2008=100)	(6) Dollar value of reduced fuel consumption (millions 2008$)
1995	1017	1.11	1128.9	75.160	1502.0
1996	1040	1.24	1289.6	76.591	1683.7
1997	1005	1.20	1206.0	77.943	1547.3
1998	1069	1.04	1111.8	78.824	1410.7
1999	1426	1.12	1597.1	79.983	1996.8
2000	1545	1.49	2302.1	81.715	2817.2
2001	1508	1.40	2111.2	83.561	2526.5
2002	1469	1.32	1939.1	84.915	2283.6
2003	1205	1.51	1819.6	86.742	2097.7
2004	1228	1.81	2222.7	89.203	2491.7
2005	1609	2.40	3861.6	92.180	4189.2
2006	1698	2.71	4601.6	95.183	4834.5
2007	1733	2.89	5008.4	97.908	5115.4
Total	17552				34496.4

Notes:
Column (4) = Column (2) x Column (3).
Column (6) = Column (4) / (Column (5) / 100).

Sources: Column (3) from EIA (2009).

Table 7.11 Emissions from 1995 through 2007

Pollutants	Reduced emissions (millions of units)
CO_2	177.3 metric tons
NO_x	0.063 tons
PM	3.808 tons
SO_x	0.096 tons

Table 7.12 Health benefits from reduced environmental emissions from the ACE R&D subprogram's research on laser and optical diagnostics and combustion modeling (rounded)

(1) Year	(2) Reduced fuel consumption with ACE R&D subprogram's technologies (million gallons)	(3) PM (g/hp-hr) per EPA regulations	(4) NO_x (g/hp-hr) per EPA regulations	(5) SO_x (ppm) per EPA regulations	(6) Monetary value of health impacts (millions 2008$)
1995	1017	0.1	5.0	500	2597.8
1996	1040	0.1	5.0	500	2681.1
1997	1005	0.1	5.0	500	2615.8
1998	1069	0.1	4.0	500	2435.4
1999	1426	0.1	4.0	500	3278.1
2000	1545	0.1	4.0	500	3675.1
2001	1508	0.1	4.0	500	3623.5
2002	1469	0.1	2.5	500	2735.7
2003	1205	0.1	2.5	500	2263.4
2004	1228	0.1	2.5	500	2327.9
2005	1609	0.1	2.5	500	3078.0
2006	1698	0.1	2.5	500	3279.0
2007	1733	0.01	1.2	500	1114.0
Total	17552				35704.8

Source: COBRA model.

The COBRA model also produces monetary values of avoided health incidents associated with the emission reductions of NO_x, PM, and SO_x shown in Table 7.11. These monetary values are shown by year in Table 7.12. The total health benefits associated with emission reductions from research in and the application of laser and optical diagnostics and

Table 7.13 Illustration of health cost calculations from the COBRA model, year 2000

(1) Category of Health Benefit	(2) Incidence	(3) Monetary Value of Health Impacts (millions 2008$)
Mortality	531	3 373.2
Infant mortality	1	9.1
Chronic bronchitis	357	158.2
Non-fatal heart attacks	836	91.9
Respiratory hospital admissions	125	1.7
Cardio-vascular related hospital admissions	258	7.2
Acute bronchitis	883	0.38
Upper respiratory symptoms	73 899	0.24
Lower respiratory symptoms	103 473	0.20
Asthma emergency room visits	466	0.17
Minor restricted-activity days	4 383 832	26.8
Work-loss days	743 012	6.0
Total		33 675.1

Source: COBRA model.

combustion modeling are estimated to be $35.7 billion (2008$) from 1995 through 2007. This estimate (Column [6], Table 7.12) is approximately equal to the dollar value of reduced fuel consumption—$34.5 billion (Column [6], Table 7.10).

Table 7.13 illustrates the monetary health benefits calculated by the COBRA model for the year 2000, the year of the greatest monetary health benefits from Table 7.12. In that year, avoided mortality accounted for 92 per cent of total monetized health benefits.

Energy Security Benefits

Security benefits are not monetized, but they are quantified in terms of the reduction of the US's dependency on imported crude oil. As discussed above and shown in Table 7.12, the ACE R&D subprogram's research on and application of laser and optical diagnostics and combustion modeling resulted in a reduction of 17.6 billion gallons of diesel fuel by heavy-duty diesel trucks from 1995 through 2007.

Approximately 10.31 gallons of diesel fuel are refined from a barrel of crude oil (DOE, 2010b). However, other petroleum products are produced from a barrel of crude oil, including gasoline, heavy fuel oil, liquefied petroleum gases, and other distillates. According to DOE (2010b), 42 gallons of crude oil (~1 barrel) yield 44 gallons of petroleum products. Thus, if one assumes that 1 gallon of imported crude oil corresponds to 1 gallon of avoided diesel fuel, then from 1995 through 2007, 17.6 billion gallons of imported crude oil have been saved, or 417.9 million barrels of imported crude oil have been avoided.

From 1995 through 2007, the US imported about 43.08 billion barrels of crude oil (DOE, 2010a). A reduction of 417.9 million barrels (0.4179 billion barrels) is approximately equal to a reduction of 1 per cent of the total crude oil imported by the US over that period of time.

ECONOMIC EVALUATION ANALYSIS

Table 7.14 shows the annual costs of EERE's investments in the ACE R&D subprogram and related CRF research costs, and the calculated annual benefits associated with the ACE R&D subprogram's research in laser and optical diagnostics and combustion modeling. Costs are relevant from 1986 through 2007,[16] and benefits are relevant from 1995 through 2007. All values are in 2008 dollars. Benefits were truncated at the end of 2007 because of lack of mileage and fuel-use data for 2008; these variables were not extrapolated to 2008 because the economy was in a recession; thus, any choice of a base year for the extrapolation would have been speculative. However, benefits beyond 2007 will continue to accrue, although they are not considered in the economic analysis, thus adding to its conservativeness.

Three economic evaluation metrics are calculated using the data in Table 7.15: present value of net benefits (net present value, *NPV*), benefit-to-cost ratio (*B*/*C*) and internal rate of return (*i*). Mathematically:

$$(NPV = \Sigma\, [B_{1995}/(1+r)^{10} + \ldots + B_{2007}/(1+r)^{22}] - \Sigma\, [C_{1986}/(1+r)^{0} + \ldots + C_{2007}/(1+r)^{21}] \tag{7.3}$$

$$B/C = \Sigma\, [B_{1995}/(1+r)^{10} + \ldots + B_{2007}/(1+r)^{22}] \,/\, \Sigma\, [C_{1986}/(1+r)^{0} + \ldots + C_{2007}/(1+r)^{21}] \tag{7.4}$$

$$\Sigma\, [B_{1995}/(1+i)^{10} + \ldots + B_{2007}/(1+i)^{22}] = \Sigma\, [C_{1986}/(1+i)^{0} + \ldots + C_{2007}/(1+i)^{21}] \tag{7.5}$$

Table 7.14 ACE R&D subprogram and CRF costs and economic benefits associated with the ACE R&D subprogram's research in laser and optical diagnostics and combustion modeling

(1) Year	(2) Costs: ACE R&D subprogram (millions 2008$)	(3) Costs: CRF (millions 2008$)	(4) Total costs (millions 2008$)	(5) Total economic benefits (millions 2008$)
1986	27.402	5.602	33.004	–
1987	29.005	5.930	34.935	–
1988	27.785	5.680	33.465	–
1989	26.525	5.423	31.948	–
1990	25.929	5.588	31.517	–
1991	22.869	6.240	29.109	–
1992	23.611	6.223	29.834	–
1993	20.550	6.073	26.623	–
1994	17.587	5.665	23.252	–
1995	13.890	5.549	19.439	4099.8
1996	21.574	6.154	27.728	4364.8
1997	24.714	6.743	31.457	4163.1
1998	23.239	6.547	29.786	3846.1
1999	46.230	6.281	52.511	5274.9
2000	57.211	5.796	63.007	6492.3
2001	62.475	6.538	69.013	6150.0
2002	55.538	6.332	61.870	5019.3
2003	63.714	6.842	70.556	4361.1
2004	59.119	6.605	65.724	4819.6
2005	52.593	6.983	59.576	7267.2
2006	42.649	6.567	49.216	8113.5
2007	49.379	7.811	57.190	6229.4
Total	793.59	137.17	930.76	70201.1

Notes:

Cost data available through 2008 in Table 3.1, but benefits data only available through 2007.

Column (5) = Column (6) in Table 7.10 + Column (6) in Table 7.12.

where, in equations (7.3) through (7.5), B represents annual total economic benefits from Column (5) in Table 7.14 and C represents total research costs from column (4) in Table 7.15. In equations (7.3) and (7.4), r is the discount rate used to reference previous years' benefits and cost to the beginning of 1986.[17, 18] In equation (7.5), i is the internal rate of return equal to that rate that equates the present value of benefits to the present value of costs.

Two alternative discount rates, r, are used in the economic evaluation.

Table 7.15 Evaluation metrics calculated from the cost and benefit data in Table 7.14

Metric	7% discount rate	3% discount rate	Internal rate of return
Present value of net benefits (billions 2008$)	$23.1	$42.6	
Benefit-to-cost ratio	53 to 1	66 to 1	
Internal rate of return			63%

The first equals the real, inflation-adjusted rate of 7 per cent (OMB, 1992),[19] and the second equals the real, inflation-adjusted rate of 3 per cent (OMB, 2003).

The values of the three economic evaluation metrics are provided in Table 3.15. The present value of net benefits is equal to $23.1 billion (using a 7 per cent discount rate), the BCR is 53 to 1, and the internal rate of return is 63 per cent. The net economic benefits of DOE-funded research on laser and optical diagnostic technologies suggest that this use of public money has been socially valuable.

CONCLUSIONS

The analysis in this chapter shows that EERE's research investments in the ACE R&D subprogram's research in and application of laser and optical diagnostics and combustion modeling and supporting DOE investments in the CRF have been socially valuable. This conclusion follows from the calculation of three traditional economic evaluation metrics: present value of net benefits ($23.1 billion, at a 7 per cent discount rate), BCR (53 to 1, at a 7 per cent discount rate), and the IRR (63 per cent).

These metrics are calculated on the basis of extant public-domain data and a set of operational assumptions, which are as follows:

- In the absence of the ACE R&D subprogram's research in and application of laser and optical diagnostics and combustion modeling to heavy-duty diesel engines, BTE from 1995 through 2007 would have been 4.5 per cent lower per year than it actually was over that period of time.
- In the absence of the ACE R&D subprogram's research in and application of laser and optical diagnostics and combustion modeling, the US diesel engine industry would not have been able to

conduct the research necessary to duplicate these technologies to heavy-duty diesel engines, even with the research assistance of universities.

- The statistical relationship between BTE and MPG from 1970 to 2007 shown in Table 7.7 applies, on average, to the intra-year statistical relationship between BTE and MPG from 1995 through 2007.

The first assumption about BTE is based on interview data from scientists at each of three companies—Caterpillar, Cummins Engine, and Detroit Diesel. On the one hand, three scientists is a small sample from which to gather such a critical piece of information, although there was consistency among the three that the reduction in BTE would have been between 4 and 5 per cent per year. On the other hand, these scientists have a unique insight to provide information about the effect of the ACE R&D subprogram's research.

The second assumption about attribution follows from three independent sources: a documented argument used to justify the initial creation of the CRF, insight from DOE scientists about industry's research capabilities, and economic theory.

The third assumption about the interpretation of the regression relationship between BTE and MPG results from limited data, although to the extent possible, the assumption was statistically verified through alternative specification tests.

As a final word of caution, based on the findings of this case study one should not generalize about the net benefits from EERE's investments in other subprograms within the VTP or within other energy programs.

NOTES

1. Between 1970 and 2007, the average annual percentage increase in truck registrations was 2.5 per cent; the average annual increase in miles driven was 3.9 per cent.
2. $$\eta_{brake} = \frac{P}{\dot{m}_f Q_{hv}}$$
   where
   η_{brake} = brake thermal efficiency
   P = engine power output measure, for example, on a dynamometer (in our unit system, this is horsepower)
   $\dot{m}_f$ = mass flow rate of fuel into the engine (mass per unit of time)
   Q_{hv} = heating value of the fuel (the amount of energy per unit of mass).
   The numerator is the power out of an engine and the denominator is the energy flowing into the engine in the form of chemical energy in the fuel.
3. This relationship is about one to one. If the BTE of new engines increases from 0.40 to 0.44 in year t, which is a 10 per cent increase, there will be a decrease of about 10 per

cent in the fuel consumption of new engines in year *t* and an increase of about 10 per cent in miles per gallon of new engines in year *t*.

4. Contacts were provided by Singh (September 9, 2009). It was agreed with each industry expert during his telephone interview that his name would not be reported in this study and that specific responses would not be attributed to his company. On the one hand, it could be argued that scientists at these companies, all of which have been funded by the DOE in the past, might be less than objective when providing information related to this study. On the other hand, no other companies have the insight necessary to provide the information needed in the calculation of benefits for this study (or that would be likely to participate in such a study).
5. This information is based on market-share information provided by one of these companies. Caterpillar is in the process of exiting the heavy-duty truck market, and as their market share declined in 2008 and 2009, the market share of Cummins and Detroit Diesel increased.
6. This statement implies that there was a ten-year lag between laser and optical diagnostics and combustion modeling research relevant to heavy-duty diesel engines and its impact on newly manufactured engines. According to McLean (October 22, 2009), a ten-year lag was relevant then but not now. Today, primarily because of more powerful lasers and improved measurement tools, the lag between ACE R&D's research and its application in industry is much shorter, possibly a few years for incremental improvements in efficiency.
7. The initial response by company scientists about the date or time period when laser and optical diagnostics and combustion modeling began to have an impact on the BTE of heavy-duty diesel engines was 'mid 1990s.' When queried about an exact impact date, 1995 was the year that was always given. Still, 1995 should be viewed as a point estimate.
8. These starting dates for the relevant research, and hence for cost, came from Siebers (October 13, 2009) and McLean (October 22, 2009).
9. Responses from all three company scientists were the same: 4 to 5 per cent. The mean response was 4.5 per cent. Ample interview time was spent with company scientists to ensure that each fully understood the counterfactual interview question about the post-1995 trend in BTE absent the ACE R&D subprogram's diagnostics and modeling research.
10. Also mentioned by the industry scientists was that EERE's investments in the entire ACE R&D subprogram have increased the fuel efficiency of heavy-duty diesel engines by 5–10 per cent since 1995. No analysis was done on this reported fuel saving.
11. All of the company scientists were of the opinion that a vertical drop in Figure 7.1 of 4.5 per cent of the graphed value beginning in 1995 is the most appropriate characterization of the counterfactual scenario. None were of the opinion that BTE would, instead, drop gradually over time. In fact, one scientist drew on a rendition of Figure 7.1 his view of the counterfactual BTE diagram to emphasize his opinion about a vertical drop of 4.5 per cent beginning in 1995.
12. One industry interviewee emphasized the importance of the public/private partnership among the DOE laboratories, industries, and universities (participating through DOE funding). Absent this public/private partnership, many, if not most, of the advancements in heavy-duty engine efficiency would not have occurred. Relatedly, Aneja and Kayes (2009) and DOE (2006), using the pre-2002 data in Figure 7.1, projected that BTE would have continued to decline after 2002 if DOE funding had not continued. In fact, this scenario was presented by them at the 2009 Directions in Engine-Efficiency and Emissions Research conference.
13. Factors not in the model that are positively related to MPG are truck weight, aerodynamic design of the truck, and improved tires. Data on these variables for heavy-duty diesel trucks were not readily available.
14. This calculation underlies the BTE Absent ACE R&D Research on Diagnostics and Modeling curve in Figure 7.2.

15. These calculations implicitly assume that the average relationship between BTE and MPG from 1970 through 2007 also holds for intra year periods from 1995 through 2007. To test this assumption empirically, equation (7.1) was reestimated with three additional regressors, each constructed as the product of BTE_t times the relevant binary variable for each period of EPA regulation. As a group, the regulation period-specific coefficients on BTE_t were not statistically different than zero.
16. In concept, an additional cost category could be considered in an economic evaluation of EERE's investments in the ACE R&D subprogram's research in laser and optical diagnostics and combustion modeling. This additional cost category relates to the investment cost that industry incurred to pull in the developed technologies and apply them to manufacturing processes. The justification for this comes from three sources: (1) discussion with industry scientists involved in diesel engine combustion; (2) the documented fact that it was very common to have one or more scientists from industry in residence at the CRF and participating in the research, often on cylinders or engines brought to the CRF by visiting researchers and then left for the CRF to use (Carlisle et al., 2002); and (3) Gunn (6 November 2009). These costs are assumed to be zero for lack of more specific information, although Gunn did speculate that in some years, industry's 'donated' equipment could have had a value equal to as much as 5 per cent of the CRF's annual budget.
17. Because all of the benefit and cost values are in 2008 dollars, a real, inflation-adjusted discount rate is appropriate.
18. Costs are assumed to be incurred at the beginning of each year, but benefits are assumed to be realized at the end of each year. Thus, the time period for the discounting of B is a year longer than for C.
19. Fundamental to implementing the present value of net benefits or the BCR is a value for the discount rate, r. Although the discount rate representing the opportunity cost for public funds could differ across a portfolio of public investments available to the DOE, the evaluation metrics in this study follow the guidelines set forth by OMB (1992) in Circular A-94, under the authority of the Budget and Accounting Act of 1921: 'Constant-dollar benefit-cost analyses of proposed investments and regulations should report net present value and other outcomes determined using a real discount rate of 7 percent.'

8. Conclusions

Escalating energy demand may be the most important issue facing the US and the world today. The production, processing, distribution, and conversion of energy consume an increasingly large share of economic activity; they drive international politics and they threaten the long-term health of the planet. As a result, there is little disagreement that R&D is needed to develop the energy technologies of the future. However, there is less agreement over the exact research agenda to be pursued, the timing of related R&D activities, and how those activities should be funded.

A key question for the US and other market-based countries is determining the roles of the private sector versus the public sector when setting the agenda for energy R&D. It is generally accepted that basic R&D has public good characteristics and that there is a strong role for government guidance and subsidies. Similarly, in the US and other market-based countries it is generally accepted that commercialization should be left primarily to the private sector because governments are not always best suited to select the winners in emerging technology markets. However, this leaves a sizable gray area in the middle where it is unclear, and case specific, as to the roles the private sector and public sector should play in developing and deploying new technologies to maximize social well-being. Applied R&D, development of generic and infratechnologies, and demonstration of the feasibility of high-risk, high-return technologies are examples of such gray areas.

In light of these uncertainties, it is important to have the proven analysis tools, methodologies, and metrics for assessing the appropriate role for government in developing and deploying new energy technologies. For example, policy makers must acknowledge that sunk costs and entrenched interests are associated with our current fossil fuel-based energy infrastructure. We need to be able to assess how this distorts what would be considered a socially optimal energy research agenda, the role government should play to address these market failures, what government investments are needed, and if past investments have been successful (i.e., are there lessons to be learned from retrospective analysis?).

This book is a first step toward investigating government's involvement in energy R&D initiatives. We investigate the need for government

Table 8.1 Benefit-cost analysis results from the case studies

	Net benefits (Billions 2008$)	Benefit-cost ratio	Internal rate of return
Solar Technology Program			
PV at 7%	1.5	1.8 to 1	17%
PV at 3%	5.7	3.2 to 1	
Geothermal Technology Program			
PV at 7%	8.1	4.9 to 1	22%
PV at 3%	17.0	9.1 to 1	
Vehicle Technology Program			
PV at 7%	23.1	53 to 1	63%
PV at 3%	42.6	66 to 1	

Note: Base year is 1976, which is the first year of DOE program expenses.

involvement in the context that the private sector would not have independently pursued these initiatives. We highlight, through examples, the need for robust evaluation methods to document social returns to taxpayers's R&D investments. The evaluation methods we consider involve developing alternative historical technology/market pathways, referred to as counterfactual scenarios, from which the benefits of government research can be measured.

This book builds on three retrospective impact assessments conducted for the DOE/EERE. These detailed economic impact evaluations found significant returns to government investments in energy technology. The benefit-cost metrics from the three case studies analyzed herein are summarized in Table 8.1. In all cases, positive net present value returns were generated with internal rates of return ranging from 17 per cent to 63 per cent.

The three studies are retrospective in that only benefits and costs through 2008 are included in the analysis. As a result, the measures of economic return presented in this book are conservative, because in many instances, the DOE's historical R&D activities will continue to generate benefits well into the future. In addition, the nature of the cluster analysis (where total program costs are compared with benefits from a subset of selected technologies) contributes to the conservative nature of the empirical findings.

Also, there are likely additional benefits associated with the three technologies that could not be quantified given the timing and resources available for the studies. For example, environmental health benefits only capture the impact of reducing emissions of PM, NO_x, and SO_x and do not reflect the benefits of reducing other pollutants such as CO_2 or mercury.

Based on the findings from these studies we summarize below several key observations that have direct implications for the US's energy R&D agenda.

MARKET FAILURES EXIST; THUS, THE PUBLIC SECTOR HAS A KEY ROLE IN SETTING AND FUNDING THE ENERGY R&D AGENDA

The theoretical basis for government's role in market activity is the economic concept of market failure. Market failure is typically attributed to market power, imperfect information, externalities, and public goods. The explicit application of market failure to justify government's role in innovation, and in R&D activity in particular, is a relatively recent phenomenon within public policy.

Pursuing next-generation energy research by its very nature involves high technical and market risk, meaning the outcomes of the firms' R&D might not be technically sufficient to meet needs. If needs are then met the market for the technology might not be mature. This leads to market failure, given that when the firm is successful, the private returns will fall short of both the firm's private hurdle rate as well as society's hurdle rate. As a result, there is an underinvestment in certain types of energy R&D from a social perspective.

High technical risk can translate to high commercial or market risk, when the requisite R&D is highly capital intensive, such as in the energy sector. As seen in the example of the development of solar technologies, many types of investment require too much capital for a single firm to justify the outlay, even though society would be better off if it had.

In addition, many R&D energy projects are characterized by a lengthy time interval until a commercial product reaches the market. The time expected to complete the R&D, and the time until commercialization of the R&D results, is long; thus, the realization of a cash flow is distant.

In addition, it is not uncommon for the scope of potential markets to be broader than the scope of the individual firm's market strategies, so the firm will not perceive economic benefits from all potential market applications of the technology. The evolving nature of markets requires investment in combinations of technologies that, if they existed, would reside in different industries that are not integrated. Because such conditions often transcend the R&D strategy of individual firms, such investments are not likely to be pursued. This was the case in the example of the GTP's development of PDC drill bits. The PDC technology developed turned out not to be suitable for geothermal applications (what it was originally designed for). However, it was rapidly adopted by the oil and gas sector, significantly reducing deep-sea drilling costs.

THE TIME FRAME FOR ENERGY TECHNOLOGY R&D PROGRAMS SHOULD REFLECT THE INCUBATION TIME REQUIRED FOR THE TECHNOLOGY

The DOE's energy R&D programs reflected the support, expertise, and long-term commitment needed to support renewable energy technologies. Our evaluations found that companies developing solar, geothermal, and vehicle technologies benefited greatly from decade-long technology development programs managed by engineers with first-hand R&D experience.

Industry believes that EERE identified technical approaches worth supporting many years before venture capital firms did, and noted that even these firms rely on EERE's independent assessments when making investment decisions. Receipt of DOE funding is viewed as a stamp of approval that the technical focus of a company is worth funding, particularly because it was a competitive procurement. The fact that DOE was willing to invest in a technology gave investors confidence that investing in the company was worthwhile. EERE performed technological diligence with a rigor of which private investors were not capable; financiers look to EERE experts for validation of a start-up's approach and for confirmation of technical claims. EERE also provided stability for programs as they sought to launch technologies that were in the nations' interest.

THE PUBLIC-SECTOR EXPERTISE MANAGING THE PROGRAM SHOULD REFLECT THE CAPABILITIES REQUIRED TO EXECUTE THE R&D

In conducting this research and engaging technologists from universities and private companies, the following became readily apparent: There is no substitute for the depth of technical expertise and service provided by EERE's divisions. Intimate technical knowledge and institutional memory were critical for program success, and industry experts strongly advocated for technology development programs to be led by EERE and its expert staff. For example, key factors in the solar technology development process were JPL leading FSA in the 1970s and 1980s, and NREL assuming PV leadership and leading programs after 1985. What is more, NREL's PVMaT and TFP program directors's technologist perspective and depth of experience in PV was cited as critical to program success.

Industry and academic experts concluded that without these programs not only would the state of energy be significantly poorer, but many US companies would not exist. The influence of the DOE and the companies it funded is reflected in the scientific literature. For solar technologies this includes factory automation for scale, encapsulants, thin-film PV, differential processing of ingots, measurement, and characterization—all of this research was enabled by DOE, which in turn drastically reduced the levelized cost of electricity, and in so doing supported demand-side policies in fueling the accumulation of installed, clean, PV-energy systems.

DIVERSITY SHOULD BE A CORNERSTONE OF AN ENERGY R&D AGENDA

Over the past 30 years, a wide range of energy technology research programs have been pursued, many of which have shown promise and to varying degrees been promoted as a the next solution to the energy problem. However, it has become increasingly clear that in the near to midterm there is not likely to be a breakthrough innovation that on its own will solve the world's energy problems.

To meet the energy challenges of the 21st century, improvements are going to be needed across a broad range of energy technologies. Research cannot just focus on a few high-profile, high-impact (but maybe low probability) energy initiatives. A viable portfolio of energy technology research is needed and should also include low-technology solutions and incremental improvements to existing technologies.

There is a role for government investment in research throughout a technology's life cycle, and to its credit, DOE has historically pursued a portfolio approach to energy technology development. However, much of the R&D public policy discussion has focused only on early stage R&D, moving new technologies from laboratories to commercialization or developing radically new energy infrastructures. The case studies in this book demonstrate that significant social returns can be achieved by removing barriers to enhancing and implementing existing technologies that are at different levels of maturity.

The key point is that in addition to the importance of aggressively pursuing R&D that will enable the next-generation energy infrastructure is the necessity to pursue near-term solutions and enhancements to existing and mature energy technologies. As seen in the example of the DOE's

VTP, relatively modest efficiency improvements in widely adopted mature technologies can lead to significant economic and environmental benefits.

EVALUATION HAS A KEY ROLE TO PLAY IN DEVELOPING, MONITORING AND MODIFYING R&D AGENDAS

Given the US's chronic budget deficits, accountability is becoming increasingly important in securing public-sector investment. In an environment of scarce resources, measurable, documented benefits are essential for setting national priorities. Impact evaluations of public-sector R&D investments should be the norm, not the exception.

Our experience indicates that economics-based evaluation methodologies are the most appropriate approach when assessing the public sector's investment costs in R&D.

The Griliches model for characterizing the benefits from a public-sector innovation has long been the cornerstone for a traditional methodology in economics for analyzing public-sector R&D programs. The Griliches model for calculating economic social rates of return adds the public and the private investments through time to determine social investment costs, and then the stream of new economic surplus generated from those investments is the benefit. Thus, the evaluative question that can be answered from such an economics-based methodology is: What is the social rate of return to the innovation, and how does it compare to the private rate of return?

In practice, the stream of new economic surplus that is generated from the public-sector investment is approximated in terms of the social benefits that exist with the public sector R&D and, counterfactually, without the public sector R&D. This so-called counterfactual approach underlies the evaluation methodology used in the three case studies presented and analyzed in this book.

Appendix: Co-Benefits Risk Assessment (COBRA) model

INTRODUCTION

The Co-Benefits Risk Assessment (COBRA) model provides estimates of health effect impacts and the economic value of these impacts resulting from emission changes. The COBRA model was developed by the US Environmental Protection Agency (EPA) to be used as a screening tool that enables users to obtain a first-order approximation of benefits due to different air pollution mitigation policies.

At the core of the COBRA model is a source-receptor (S-R) matrix that translates changes in emissions to changes in particulate matter (PM) concentrations. The changes in ambient PM concentrations are then linked to changes in mortality risk and changes in health incidents that lead to health care costs and/or lost workdays.

CHANGES IN EMISSION → CHANGES IN AMBIENT PM CONCENTRATIONS

The user provides changes (decreases) in emissions of pollutants ($PM_{2.5}$, SO_2, NO_x) and identifies the economic sector from which the emissions are being reduced. These changes are in total tons of pollutants by sector for the US economy for the chosen year of analysis. The economic sectors chosen determine the underlying spatial distribution of emissions and hence the characteristics of the human population that is affected. For example, emission reduction due to the use of geothermal technology is typically applied to coal plants in electric utilities. Reductions due to the use of wind technology are applied to coal, oil, and natural gas plants in electric utilities. Emissions reductions due to improved efficiency of diesel engines are applied to both highway diesel engines and off-highway nonroad diesel engines.

The S-R matrix consists of fixed transfer coefficients that reflect the relationship between annual average $PM_{2.5}$ concentration values at a single receptor in each county (a hypothetical monitor located at the county centroid) and the contribution by $PM_{2.5}$ species to this concentration from each emission source. This matrix provides quick but rough estimates of the impact of emission changes on ambient $PM_{2.5}$ levels as compared with the detailed estimates provided by more sophisticated air-quality models (EPA, 2006).

CHANGES IN AMBIENT PM CONCENTRATIONS → CHANGES IN HEALTH EFFECTS

The model then translates the changes in ambient PM concentration to changes in incidence of human health effects using a range of health impact functions and estimated baseline incidence rates for each health endpoint. The data used to estimate baseline incidence rates and the health impact functions used vary across the different health endpoints. To be consistent with prior EPA analyses, the health impact functions and the unit economic value used in COBRA are the same as the ones used for the Regulatory Impact Analysis of the Clean Air Interstate Rule (EPA, 2005).

The model provides (in the form of a table or map) changes in the number of cases for each health effect between the baseline emissions scenario (included in the model) and the analysis scenario. The different health endpoints are included in Table A.1.

Each health effect is described briefly below. For additional detail on the epidemiological studies, functional forms and coefficients used in COBRA, see Appendix C of the COBRA user's manual (EPA, 2006; Abt, 2009).

Mortality researchers have linked both short-term and long-term exposure to ambient levels of air pollution to increased risk of premature mortality. COBRA uses mortality risk estimates from an epidemiological study of the American Cancer Society cohort conducted by Pope et al. (2002). COBRA includes different mortality risk estimates for both adults and infants. Because of the high monetary value associated with prolonging life, mortality risk reduction is consistently the largest health endpoint valued in the study.

Chronic bronchitis is defined as a persistent wet cough and mucus in the lungs for at least three months for several consecutive years. It affects approximately 5 per cent of the population (Abt, 2009). A study by Abbey et al. (1995) found statistically significant relationships between $PM_{2.5}$ and PM_{10} and chronic bronchitis.

Peters et al. (2001) linked nonfatal heart attacks to PM exposure. These heart attacks are modeled separately from hospital admissions because of their lasting impact on long-term health care costs and earning.

Table A.1 Health endpoints included in COBRA

Health effect	Description
Mortality	Number of deaths
Chronic bronchitis	Cases of chronic bronchitis
Non-fatal heart attacks	Number of non-fatal heart attacks
Respiratory hospital admissions	Number of cardiopulmonary-, asthma-, or pneumonia-related hospitalizations
Cardio-vascular related hospital admissions	Number of cardiovascular-related hospitalizations
Acute bronchitis	Cases of acute bronchitis
Upper respiratory symptoms	Episodes of upper respiratory symptoms (runny or stuffy nose; wet cough; and burning, aching, or red eyes)
Lower respiratory symptoms	Episodes of lower respiratory symptoms: cough, chest pain, phlegm, or wheeze
Asthma emergency room visits	Number of asthma-related emergency room visits
Minor restricted-activity days	Number of minor restricted-activity days (days on which activity is reduced but not severely restricted; missing work or being confined to bed is too severe to be MRAD)
Work-loss days	Number of work days lost due to illness

Hospital admissions consist of two major categories: respiratory (such as pneumonia and asthma) and cardiovascular (such as heart failure, ischemic heart disease). Using detailed hospital admission and discharge records, Sheppard et al. (1999) investigated asthma hospital emissions associated with PM, carbon monoxide (CO), and ozone, and Moolgavkar (2000 and 2003) and Ito (2003) found a relationship between hospital admissions and PM. COBRA includes separate risk factors for hospital admissions for people aged 18 to 64 and aged 65 or older.

Acute bronchitis, defined as coughing, chest discomfort, slight fever, and extreme tiredness lasting for a number of days, was found by Dockery et al. (1996) to be related to sulfates and particulate acidity and, to a lesser extent, PM. COBRA estimates the episodes of acute bronchitis in children aged 8 to 12 from pollution using the findings from Dockery et al.

Upper respiratory symptoms include episodes of a runny or stuffy nose, a wet cough, and burning, aching, or red eyes. Pope et al. (2002) found a relationship between PM and the incidence of these same symptoms.

Lower respiratory symptoms in COBRA are based on Schwarz and Neas (2000) and focus primarily on children's exposure to pollution.

Children were selected for the study based on indoor exposure to PM and other pollutants resulting from parental smoking and gas stoves. Episodes of lower respiratory symptoms are coughing, chest pain, phlegm, or wheezing.

Asthma-related emergency room visits are primarily associated with children under the age of 18. Norris et al. (1999) found significant associations between these visits and PM and CO. To avoid double counting, hospitalization costs (discussed above) do not include the cost of admission to the emergency room.

Ostro and Rothschild (1989) researched minor restricted-activity days (MRAD) in COBRA. MRADs include days on which activity is reduced but not severely restricted (e.g., missing work or being confined to bed is too severe to be an MRAD). They estimated the incidence of MRADs for a national sample of the adult working population, aged 18 to 65, in metropolitan areas. Because this study is based on a convenience sample of nonelderly individuals, the impacts may be underestimated because the elderly are likely to be more susceptible to PM-related MRADs.

Work-loss days were estimated by Ostro (1987) to be related to PM levels. Based on an annual national survey of people aged 18 to 65, Ostro found that two-week average PM levels were significantly linked to work-loss days. However, the findings showed some variety across years.

CHANGES IN HEALTH EFFECTS → CHANGES IN MONETARY IMPACTS

COBRA translates the above-listed health effects into changes in monetary impacts using estimated unit values of each health endpoint. The per-unit monetary values are described in Appendix F of the COBRA user's manual (EPA, 2006). Estimation of the monetary unit values varies by the type of health effect. For example, reductions in the risk of premature mortality are monetized using value of statistical life (VSL) estimates. Other endpoints such as hospital admissions use cost of illness (COI) units that include the hospital costs and lost wages of the individual but do not capture the social (personal) value of pain and suffering.

LIMITATIONS

It should be noted that COBRA does not incorporate effects of many pollutants, such as carbon emissions or mercury. This has two potential implications. First, other pollutants may cause or exacerbate health endpoints that are not included in COBRA. This would imply that reducing incidences of such health points are not captured. Second, pollutants other than those included in COBRA may also cause a higher number of incidences of the health effects that are part of the model. This is also not captured in this analysis. Thus, the economic value of health effects obtained from COBRA may be interpreted as a conservative estimate of the health benefits from reducing emissions.

References

Abbey, D.E., B.E. Ostro, F. Petersen et al. (1995), 'Chronic respiratory symptoms associated with estimated long-term ambient concentrations of fine particulates less than 2.5 microns in aerodynamic diameter ($PM_{2.5}$) and other air pollutants,' *Journal of Exposure Analysis and Environmental Epidemiology*, **5** (2), 137–59.

Abt Associates Inc. (2009), 'Economic impact of Wise County, Virginia Coal-Fired Power Plant,' prepared for Wise Energy For Virginia Coalition, January.

AIAA (American Institute of Aeronautics and Astronautics) (2009), 'Laser-induced incandescence,' available at: https://info.aiaa.org/tac/ASG/AMTTC/Shared%20Documents/Technique%20Overviews/lii.html (accessed 23 September 2009).

Allen, M.G. (1998), 'Diode laser absorption sensors for gas-dynamic and combustion flows,' *Measurement Science and Technology*, **9** (4), 545–62, available at: http://www.ncbi.nlm.nih.gov/pubmed/11543363 (accessed 28 July 2011).

American Presidency Project (1999–2011), 'Richard Nixon: address on the State of the Union delivered before a joint session of the Congress,' available at: http://www.presidency.ucsb.edu/ws/index.php?pid=4327#axzz1TzcTPc66 (accessed 9 August 2011).

Amsden, D.C. and A.A. Amsden (1993), 'The KIVA story: a paradigm of technology transfer,' *IEEE Transactions on Professional Communication*, **36**, 190–95.

Aneja, R., Y. Kalish, and D. Kayes (2009), 'Integrated powertrain and vehicle technologies for fuel efficiency improvement and CO_2 reduction,' paper presented at the Directions in Engine-Efficiency and Emissions Research Conference (DEER), August 5.

Aneja, R. and D. Kayes (2009), 'Reduction of heavy-duty fuel consumption and CO_2 generations: what the industry does and what government can do,' paper presented at the Directions in Engine-Efficiency and Emissions Research Conference (DEER), August 5.

Asanuma, T. (1996), 'New visualization and image techniques for engine combustion research,' in the Visualization Society of Japan (ed.), *Atlas of Visualization*, Boca Raton, FL, USA CRC Press, pp. 3–41.

Blankenship, D. (2009), Personal communication between Michael Gallaher and Alex Rogozhin (RTI International), and Doug Blankenship (Sandia National Laboratory).

Boudreaux, R. and K. Massey (1994), 'Turbodrills and innovative PDC bits economically drilled hard formations,' *Oil and Gas Journal*, **92** (13), 52–5.

Bresnahan, T.F. and M. Trajtenberg (1995), 'General purpose technologies,' *Journal of Econometrics*, **65**, 83–108.

Carlisle, R.P., D.J. Monetta, and W.L. Sparks (2002), *The Combustion Research Facility: Model for a 21st-Century Open User Facility*, Albuquerque, NM, USA and Livermore, CA, USA: Sandia National Laboratories.

Casto, R. (1995), 'Use of bicenter PDC bit reduces drilling cost,' *Oil and Gas Journal*, **93** (46), 92–6.

CATF (Clean Air Task Force) (2005), Diesel engines: emissions sources and regulations.

Christensen, E. (1985), 'Flat-plate solar array project: 10 years of progress,' paper prepared for the US Department of Energy by the Jet Propulsion Laboratory, National Aeronautics and Space Administration.

CRS (Congressional Research Service) (2001), 'Diesel fuel and engines: an analysis of EPA's new regulations,' report to Congress, 1 May.

Davis, S.C., S.W. Diegel, and R.G. Boundy (2009), *Transportation Energy Data Book*, 28th edition, Oak Ridge, TN, USA: Oak Ridge National Laboratory.

DoC (Department of Commerce) (2009), 'National income and product accounts table: table 1.1.9. implicit price deflators for gross domestic product,' revised October 29, 2009. Bureau of Economic Analysis, available at: http://www.bea.gov/national/nipaweb/ (accessed 28 July 2011).

Dockery, D.W., J. Cunningham, A.I. Damokosh et al. (1996), 'Health effects of acid aerosols on North American children—respiratory symptoms,' *Environmental Health Perspectives*, **104** (5), 500–505.

DOE (Department of Energy) (1993), 'Drilling sideways: a review of horizontal well technology and its domestic application,' DOE/EIA-TR-0565, Washington, DC: Department of Energy, Energy Information Administration, available at http://tonto.eia.doe.gov/ftproot/petroleum/tr0565.pdf (accessed 28 July 2011).

DOE (2003), 'Just the basics: diesel engine,' Washington DC: Office of Energy Efficiency and Renewable Energy, available at: http://www1.eere.energy.gov/vehiclesandfuels/pdfs/basics/jtb_diesel_engine.pdf (accessed 28 July 2011).

DOE (2006), '21st-century truck partnership: roadmap and technical white papers,' 21CTP-0003, available at: http://www1.eere.energy.gov/vehiclesandfuels/pdfs/program/21ctp_roadmap_2007.pdf (accessed 28 July 2011).

DOE (2008a), 'DOE solar energy technologies program. FY 2007 annual report,' available at: http://www.nrel.gov/docs/fy08osti/42549.pdf (accessed 28 July 2011).

DOE (2008b), 'FY 2008 progress report for advanced combustion engine technologies,' Washington DC: Office of Energy Efficiency and Renewable Energy, Vehicle Technologies Program, available at: http://www1.eere.energy.gov/vehiclesandfuels/pdfs/program/2008_adv_combustion_engine.pdf (accessed 28 July 2011).

DOE (2008c), 'A history of geothermal energy R&D in the United States: geoPowering the West,' Draft paper, Washington, DC: US Department of Energy.

DOE (2009a), 'DOE solar energy technologies program. FY 2008 annual report,' available at: http://www.nrel.gov/docs/fy09osti/43987.pdf (accessed 28 July 2011).

DOE (2009b), 'Historical savings,' presentation of cost and energy savings data provided to the Office of Energy Efficiency and Renewable Energy by Caterpillar, Cummins, and Detroit Diesel, September 10.

DOE (2010a), 'US imports of crude oil,' Energy Information Administration, available at: http://tonto.eia.doe.gov/dnav/pet/hist/LeafHandler.ashx?n=PET&s=MCRIMUS1&f=M (accessed 8 February 2010).

DOE (2010b), 'Oil crude and petroleum products explained,' Energy Information Administration, available at: http://tonto.eia.doe.gov/energyexplained/index.cfm?page=oil_home (accessed 8 February 2010).

DOE (2010c), 'Geothermal Technologies Program: About the Program,' Washington, DC: US DOE, Energy Information Administration, available at: http://www1.eere.energy.gov/geothermal/about.html (accessed 28 July 2011).

DOE (2010d), Table 8.2a: Electricity net generation, 1949–2008, available at http://www.eia.doe.gov/emeu/aer/txt/stb0802a.xls (accessed on 9 August 2011).

DOE (2011), Mission, available at: http://energy.gov/mission (accessed 9 August 2011).

Dooley, J.J. (2008), 'US federal investments in energy R&D: 1961–2008,' Pacific Northwest National Laboratory, US Department of Energy, October.

Eberhardt, J.J. (2009a), Personal interview, 26 June.

Eberhardt, J.J. (2009b), E-mail correspondence, 31 July.

Eberhardt, J.J. (2009c), Telephone interview, 9 September.

Eberhardt, J.J. (2010), E-mail correspondence, 20 January.

EERE (2009a), 'Vehicle Technologies Program: program areas,' available at: http://www1.eere.energy.gov/vehiclesandfuels/program_areas/index.html (accessed 9 November 2009).

EERE (2009b), 'PV panel disposal and recycling,' Washington, DC: US Department of Energy.

EERE/GTP (Energy Efficiency and Renewable Energy/Geothermal Technologies Program) (2008a), 'A history of geothermal energy research and development: drilling 1976 to 2006,' Final Lab Draft, Washington DC: US Department of Energy, 31 December.

EERE/GTP (2008b), 'A history of geothermal energy research and development: exploration 1976 to 2006,' Final Lab Draft, Washington DC: US Department of Energy, 31 December.

EERE/GTP (2008c), 'A history of geothermal energy research and development: reservoir engineering 1976 to 2006,' Final Lab Draft, Washington DC: US Department of Energy, 31 December.

EERE/GTP (2008d), 'A history of geothermal energy research and development: energy conversion 1976 to 2006,' Final Lab Draft, Washington DC: US Department of Energy, 31 December.

EERE/GTP (2010), 'A history of geothermal energy in the United States,' available at: http://www1.eere.energy.gov/geothermal/history.html (accessed 28 July 2011).

EIA (Energy Information Administration) (2008), 'Annual photovoltaic module/cell manufacturers survey,' Washington, DC: US Department of Energy, available at: http://www.eia.doe.gov/cneaf/solar.renewables/page/solarphotv/solarpv.html (accessed 28 July 2011).

EIA (2009a), 'Annual energy review: petroleum,' Washington DC: US Department of Energy, available at: http://www.eia.DOE.gov/emeu/aer/petro.html (accessed 21 October 2009).

EIA (2009b), 'Appendix B: emissions factors,' available at: http://www.eia.doe.gov/pub/oiaf/1605/cdrom/pdf/gg-app-tables.pdf (accessed 28 July 2011).

EIA (2009c), 'US electric power industry net generation,' Electric Power Annual with data for 2007, Washington, DC: US Department of Energy, 20 January.

EIA (2011), 'Annual energy review,' available at: http://www.eia.doe.gov/aer/txt/stb0103.xls (accessed 29 March 2011).

EPA (Environmental Protection Agency) (2005), 'Regulatory impact analysis for the final clean air interstate rule. EPA-452/R-05-002,' Research Triangle Park, NC, USA: Office of Air Quality Planning and Standards; Emission, Monitoring, and Analysis Division and Clean

Air Markets Division, available at: http://www.epa.gov/cair/pdfs/finaltech08.pdf (accessed 28 July 2011).

EPA (2006), 'User's manual for the Co-Benefits Risk Assessment (COBRA) screening model,' developed by Abt Associates Inc.

EPA (2009a), 'eGRID,' available at: http://www.epa.gov/cleanenergy/energy-resources/egrid/index.html (accessed 28 July 2011).

EPA (2009b), 'Greenhouse gas emissions,' available at: http://www.epa.gov/climatechange/emissions/index.html (accessed 28 July 2011).

EPA (2009c), 'Greenhouse gas equivalency calculator,' available at: http://www.epa.gov/cleanenergy/energy-resources/calculator.html (accessed 9 August 2011).

Executive Office of the President (1990), 'US technology policy,' Washington, DC: Office of Science and Technology Policy.

Executive Office of the President (1994), 'Economic report of the President,' Washington, DC: Government Printing Office.

Executive Office of the President (2000), 'Economic report of the President,' Washington, DC: Government Printing Office.

Falcone, S., and D. Bjornstad (2005), 'Leveraging defense research: social impact of the transfer of polycrystalline diamond drill bit research,' *Comparative Technology Transfer and Society*, **3** (December) 267–300, doi: 10.1353/ctt.2006.0004.

Fehner, T.R. and J.M. Holl (1994), *Department of Energy 1977–1994: A Summary History*, Washington, DC: US Department of Energy, History Division.

Finger, J., and D. Glowka (1989), 'PDC Bit Research at Sandia National Laboratories,' Sandia Corporation, Washington, DC: Department of Energy, available at: http://www.osti.gov/geothermal/servlets/purl/6107005-GL9XS0/6107005.pdf (accessed 28 July 2011).

Flynn, H., and T. Bradford (2006), *Polysilicon: Supply, Demand and Implications for the PV Industry*, Prometheus Institute for Sustainable Development.

Forch, B.E., J.B. Morris, and A.M. Miziolek (1990), 'Laser-induced florescence and ionization techniques for combustion diagnostics,' in T. Vo-dirh and D. Eastwood (eds), *Laser Techniques in Luminescence Spectroscopy*, Philadelphia: American Society for Testing Materials, pp. 50–68.

Fort, L.N., A.F. Burton, E.P. Bellevue et al. (1980), '*US Patent 4,217,862, High Constant Pressure, Electronically Controlled Diesel Fuel Injection System*, Washington, DC: US Patent and Trademark Office.

Freedonia Group (2009), *Drilling Products and Services to 2012. Drill Bit and Realer Demand by Product: Fixed Cutter Drill Bits*, Cleveland, OH, USA: The Freedonia Group, Inc., available at: http://www.freedonia

group.com/FractionalDetails.aspx?DocumentId=405397 (accessed 28 July 2011).

Friedman, D.J., R.L. Mitchell, B.M. Keyes, et al. (2005), *PV Manufacturing R&D Project Status and Accomplishments Under Inline Diagnostics and Intelligent Processing and Yield, Durability, and Reliability*, Golden, CO, USA: National Renewable Energy Laboratory.

Gallagher, G., P. Alexander, and D. Burger (1986), 'Volume V: Process development. Flat-Plate Solar Array Project final report,' project-managed by the Jet Propulsion Laboratory for the US Department of Energy, JPL Publication 86–31, Pasadena, CA: Jet Propulsion Laboratory.

Gani, R. (1982), 'Diamond shear bits pass test in Indonesia's Arun gas Field,' *Oil and Gas Journal*, **80** (48), 79–81.

GEA (Geothermal Energy Association) (2009), Listing of US Geothermal Power Plants.

Glitnir Geothermal Research (2008), *United States Geothermal Energy Market Report*, Reykjavik, Iceland.

Glowka, D. and D. Schaefer (1993), *Program Plan for the Development of Advanced Synthetic-Diamond Drill Bits for Hard-Rock Drilling*, Washington, DC: Department of Energy. doi: 10.2172/10104670.

Glowka, D., T. Dennis, P. Le, et al. (1995), 'Progress in the advanced synthetic-diamond drill bit program,' Washington, DC: Department of Energy, available at: http://www.osti.gov/bridge/servlets/purl/127986-TvmMq5/webviewable/127986.pdf (accessed 28 July 2011).

Green, M. (2005), 'Silicon photovoltaic modules: a brief history of the first 50 years,' *Progress in Photovoltaics: Research and Applications*, **13**, 447–55.

Green, M. (2009), 'The path to 25% silicon solar cell efficiency: history of silicon cell evolution,' *Progress in Photovoltaics: Research and Applications*, **17**, 183–9.

Griliches, Z. (1958), 'Research costs and social returns: hybrid corn and related innovations,' *Journal of Political Economy*, **66**, 419–31.

Gunn, M. (2009a), E-mail correspondence, 20 October.

Gunn, M. (200b), Telephone interview, 6 November.

Hanson, R.K., D.S. Baer, and J.B. Jeffries (2002), 'Multiplexed diode-laser absorption sensors for real-time measurements and controls of combustion systems,' final report to the National Center for Environmental Research.

Hartley, D.L. and T.M. Dyer (1985), 'New diagnostic techniques in engine combustion research,' paper presented at COMODIA 1985, the International Symposium on Diagnostics and Modeling in Internal Combustion Engines, Tokyo, Japan, 4 September.

Hertel, T.W., W.E. Tyner and D.K. Birur (2010), 'Global impacts of biofuels,' *Energy Journal*, **31** (1), 75–100.

Hulstorm, R. (2010), Personal communication with Alan C. O'Connor (RTI International), 12 March.

IEA (International Energy Agency) (2009), 'National survey report of PV power applications in the United States: 2008,' paper prepared by the National Renewable Energy Laboratory for the US Department of Energy.

Ito, K. (2003), 'Associations of particulate matter components with daily mortality and morbidity in Detroit, Michigan,' in Health Effects Institute, Revised Analyses of Time-Series Studies of Air Pollution and Health, Boston, MA, USA.

Jennings, C.E., R.M. Margolis, and J.E. Bartlett (2008), 'A historical analysis of investment in solar energy technologies (2000–2007),' Technical Report NREL/TP-6A42-43602, Golden, CO: National Renewable Energy Laboratory.

Jones, P. (1988), *Oil: A Practical Guide to the Economics of World Petroleum*, Cambridge, UK: Woodhead-Faulkner.

Kalish, Y. (2009), Telephone interview, 12 October.

Komp, R. (2001), *Practical Photovoltaics: Electricity from Solar Cells*, 3rd rev edition, Ann Arbor, MI, USA: Aatec Publications.

Kurokawa, K. and O. Ikki (2001), 'The Japanese experiences with national PV system programmes,' *Solar Energy*, **70** (6), 457–66.

Lance, M.J. and C.S. Sluder (2009), 'Materials issues associated with EGR systems,' presentation at the DOE 2009 Vehicle Technologies Annual Merit Review and Peer Evaluation Meeting, 21 April.

Linden, L., D. Bottaro, J. Moskowitz, et al. (1977), 'The solar photovoltaics industry: the status and evolution of the technology and the institutions,' MIT Energy Laboratory Report MIT-EL-77-021, prepared for the US Department of Energy.

Link, A.N. and J.T. Scott (1998), *Public Accountability: Evaluating Technology-Based Institutions*, Norwell, MA, USA: Kluwer Academic Publishers.

Link, A.N. and J.T. Scott (2005), *Evaluating Public Research Institutions: The US Advanced Technology Program's Intramural Research Initiative*, London: Routledge.

Link, A.N., and J.T. Scott. 2011. *Public Goods, Public Gains: Calculating the Social Benefits of Public R&D.* Oxford: Oxford University Press.

Lutwack, R. (1986), 'Volume II: silicon material. Flat-Plate Solar Array Project final report,' Project-managed by the Jet Propulsion Laboratory for the US Department of Energy, JPL Publication 86–31, Pasadena, CA: Jet Propulsion Laboratory.

Madigan, J. and R. Caldwell (1981), 'Application for polycrystalline diamond compact bits form analysis of carbide insert and steel tooth bit performance,' *Journal of Petroleum Technology*, **33**, 1171–9.

Mansfield, E., J. Rapoport, A. Romeo, et al. (1977), 'Social and private rates of return from industrial innovations,' *Quarterly Journal of Economics*, **91**, 221–40.

Margolis, R. (2002), 'Understanding technological innovation in the energy sector: the case of photovoltaics,' doctoral dissertation, Woodrow Wilson School of Public and International Affairs, Princeton University.

Margolis, R., R. Mitchell, and K. Zweibel (2006), 'Lessons Learned From the Photovoltaic Manufacturing Technology/PV Manufacturing R&D and Thin-Film PV Partnership Projects,' Technical Report. NREL/TP-520-39780. Golden, CO: National Renewable Energy Laboratory.

Mattison, D.W., J.B. Jeffries, R.K. Hanson, et al. (2007), 'In-cylinder gas temperature and water concentration measurements in HCCI engines using a multiplexed-wavelength diode-laser system: sensor development and initial demonstration,' *Proceedings of the Combustion Institute*, **31**, 791–8.

Maycock, P. 1986–2004. *PV News*, Warrenton, VA: PV Energy Systems.

McDonald, S., and F. Felderhoff (1996) March, 'New bits, motors improve economics of slim hole horizontal wells,' *Oil and Gas Journal*, **94** (11).

McLean, W. (2009a), pp. 66–70. E-Mail Correspondence and Telephone Interviews, 22 October.

McLean, W. (2009b), E-mail correspondence and telephone interviews, 4 November.

McLean, W. (2009c), E-mail correspondence and telephone interviews, 20 November.

Mensa-Wilmot, G. (1997), 'New PDC cutters improve drilling efficiency,' *Oil and Gas Journal*, **95** (43), 64–70.

Mensa-Wilmot, G. (2003), *PDC Drill Bit Having Cutting Structure Adapted to Improve High Speed Drilling Performance*, US Patent No. 6,615,934, Washington, DC: US Patent and Trademark Office.

Mitchell, R. L. (2009), Personal communication with Alan O'Connor, 20 August.

Mitchell, R.L., C.E. Witt, and G.D. Mooney (1992), 'The Photovoltaic Manufacturing Technology Project: A Government Industry Partnership,' NREL/TP-214-4588, Golden, CO: National Renewal Energy Laboratory.

Mitchell, R.L., M. Symko-Davies, H.P. Thomas et al. (1998), '1998

PVMaT overview,' Presented at the National Center for Photovoltaics Program Review Meeting, Denver, CO, September.
Moolgavkar, S.H. (2000), 'Air pollution and hospital admissions for chronic obstructive pulmonary disease in three metropolitan areas in the United States,' *Inhalation Toxicology*, **12** Suppl 4, 75–90.
Moolgavkar, S.H. (2003), 'Air pollution and daily deaths and hospital admissions in Los Angeles and Cook Counties,' in Health Effects Institute (ed.), *Revised Analyses of Time-Series Studies of Air Pollution and Health*, Boston, MA, USA: Health Effects Institute, pp. 193–198.
NC Solar Center (2009), *Database of State Incentives for Renewables and Efficiency (DSIRE)*, Raleigh, NC: NC Solar Center and the Interstate Renewable Energy Council, available at: http://www.dsireusa.org/ (accessed 28 July 2011).
Nam, E. (2004), *Advanced Technology Vehicle Modeling in PERE*, Washington, DC: US Environmental Protection Agency.
Norris, G., S.N. YoungPong, J.Q. Koenig, et al. (1999), 'An Association Between Fine Particles and Asthma Emergency Department Visits for Children in Seattle,' *Environmental Health Perspectives*, **107** (6), 489–93.
NRC (National Research Council) (2001), *Energy Research at DOE: Was It Worth It? Energy Efficiency and Fossil Energy Research 1978 to 2000*, Washington, DC: National Academy Press.
NRC (2008), *Review of the 21st Century Truck Partnership*, Washington, DC: National Academy Press.
NREL (National Renewable Energy Laboratory) (2009a), 'Awards and honors: R&D 100 awards,' Golden, CO: National Renewable Energy Laboratory, available at: http://www.nrel.gov/awards/rd_awards.html (accessed 28 July 2011).
NREL (2009b), 'PV manufacturing R&D: partnerships,' Golden, CO: National Renewable Energy Laboratory, available at: http://www.nrel.gov/pv/pv_manufacturing/cfm/partnerships.cfm (accessed 28 July 2011).
NREL (2009c), 'PV Manufacturing R&D: Cost/Capacity Analysis,' Golden, CO: National Renewable Energy Laboratory, available at: http://www.nrel.gov/pv/pv_manufacturing/html (accessed 28 July 2011).
NREL (2009d), 'PV manufacturing R&D: history,' Golden, CO: National Renewable Energy Laboratory, available at: http://www.nrel.gov/pv/pv_manufacturing/html (accessed 28 July 2011).
Office of Energy Research, US Department of Energy (1997), 'Energy Materials Coordinating Committee (EMaCC) Fiscal Year 1996 Annual Technical Report,' DOE/ER-0715.
Office of Energy Research, US Department of Energy (1998), 'Energy Materials Coordinating Committee (EMaCC) Fiscal Year 1997 Annual Technical Report,' DOE/ER-0734.

Office of Energy Research, US Department of Energy (1999), 'Energy Materials Coordinating Committee (EMaCC) Fiscal Year 1998 Annual Technical Report,' DOE/SC-0011.

Office of Science, US Department of Energy (2000), 'Energy Materials Coordinating Committee (EMaCC) Fiscal Year 1999 Annual Technical Report,' DOE/SC-0025.

Office of Science, US Department of Energy (2001), 'Energy Materials Coordinating Committee (EMaCC) Fiscal Year 2000 Annual Technical Report,' DOE/SC-0040.

Office of Science, US Department of Energy (2002), 'Energy Materials Coordinating Committee (EMaCC) Fiscal Year 2001 Annual Technical Report,' DOE/SC-0061.

Office of Science, US Department of Energy (2003), 'Energy Materials Coordinating Committee (EMaCC) Fiscal Year 2002 Annual Technical Report,' DOE/SC-0077.

Office of Science, US Department of Energy (2004), 'Energy Materials Coordinating Committee (EMaCC) Fiscal Year 2003 Annual Technical Report,' DOE/SC-0088.

Office of the Under Secretary of Defense for Acquisition, Technology & Logistics (2011), 'Public Law 93-438: Energy Reorganization Act of 1974,' available at: http://www.acq.osd.mil/ncbdp/nm/nmbook/references/Congress/1.11%20Energy%20Reorganization%20Act%20of%20 1974.pdf (accessed 9 August 2011).

O'Sullivan, M., K. Pruess, and M. Lippmann (2001), 'State of the art of geothermal reservoir simulation,' *Geothermics*, **30**, 395–429.

OMB (1992), *Circular No. A–94: Guidelines and Discount Rates for Benefit–Cost Analysis of Federal Programs*, Washington, DC: Government Printing Office.

OMB (Office of Management and Budget) (2003), *Circular No. A–4: Regulatory Analysis*, Washington, DC: Government Printing Office.

Osterwald, C.R., and T.J. McMahon (2009), 'History of accelerated and qualification testing of terrestrial photovoltaic modules: a literature review,' *Progress in Photovoltaics: Research and Applications*, **17**, 11–33.

Ostro, B.D. (1987), 'Air pollution and morbidity revisited: a specification test,' *Journal of Environmental Economics and Management*, **14**, 87–98.

Ostro, B.D. and S. Rothschild (1989), 'Air pollution and acute respiratory morbidity—an observational study of multiple pollutants,' *Environmental Research*, **50** (2), 238–47.

Papadakis, M. and A. Link (1997), 'Measuring the unmeasurable: benefit-cost analysis for new business start-ups and scientific research transfers,' *Evaluation and Program Planning*, **20** (1), 99–102.

Peters, A., D.W. Dockery, J.E. Muller et al. (2001), 'Increased particulate air pollution and the triggering of myocardial infarction,' *Circulation*, **103** (23), 2810–5.

PNGV (Partnership for a New Generation of Vehicles) (2009), 'About partnership for a new generation of vehicles,' available at: http://www.pngv.org/main/index.php?option=com_content&task=blogcategory&id=14&Itemid=26 (accessed 3 August 2009).

Pope, C.A., R.T. Burnett, M.J. Thun et al. (2002), 'Lung cancer, cardiopulmonary mortality, and long-term exposure to fine particulate air pollution,' *Journal of the American Medical Association*, **287** (9), 1132–41.

PV News (2005–2009), *PV News*, Warrenton, VA: PV Energy Systems.

Ruegg, R. and G. Jordan (2009), 'Guidelines for Conducting Retrospective Benefit–Cost Studies,' (Internal DOE draft report (not yet released).

Schwartz, J. and L.M. Neas (2000), 'Fine particles are more strongly associated than coarse particles with acute respiratory health effects in schoolchildren,' *Epidemiology*, **11** (1), 6–10.

SEMI (Semiconductor Equipment and Materials International) (2009), Global Silicon Wafer Diameter Trends (Millions of Square Inches), San Jose, CA.

Sheppard, L., D. Levy, G. Norris et al. (1999), 'Effects of ambient air pollution on nonelderly asthma hospital admissions in Seattle, Washington, 1987–1994,' *Epidemiology*, **10** (1), 23–30.

Siebers, D. (2009), E-mail correspondence and telephone interviews, 18 August, 8 September, 13 October, and 23 October.

Singh, G. (2000), 'Additional combustion and emission control projects, heavy truck engine program and performance measures for the engine team,' presentation at the Heavy-Duty Vehicle Technology Annual Review, 12 April.

Singh, G. (2009), Telephone interview, 9 September.

Slack, J.B. and J. Wood (1982), 'Stratapax bits prove economical in Austin Chalk,' *Oil and Gas Journal*, **79**, 164–5.

Smoller, (1996), *NREL Photovoltaic Program: FY 1995 Annual Report*, NREL/TP-410-21101.

Summers, K.A. (1991), 'Annual report: photovoltaic subcontract program FY 1990,' SERI/TP-214-4135.

Summers, K.A. (1995), 'NREL photovoltaic program: FY 1994 annual report,' NREL/TP-410-7993.

Surek, T. (1992), *Overview of NREL's Photovoltaic Advanced R&D Project*, Golden, CO: National Renewable Energy Laboratory.

Swanson, R.M. (2006), 'A vision for crystalline silicon photovoltaics,' *Progress in Photovoltaics: Research and Applications*, **14**, 443–53.

Ullal, H. (2009), Personal communication with Alan C. O'Connor and Ross J. Loomis, 4 August.

US Bureau of Census (Census) (2011), 'Statistical Abstracts,' available at: http://www.census.gov/prod/www/abs/statab.html (accessed 28 July 2011).

US Bureau of Economic Analysis (BEA) (2011), 'National economic accounts,' available at: http://www.bea.gov/national/ (accessed 28 July 2011).

United States Department of the Interior (2011), 'Energy Reorganization Act 1977,' available at: www.usbr.gov/power/legislation/doeorg.pdf (accessed 9 August 2011).

United States Nuclear Regulatory Commission (2002), 'Nuclear Regulatory Legislation,' available at: http://www.nrc.gov/reading-rm/doc-collections/nuregs/staff/sr0980/ml022200075-vol1.pdf, pp. 2–6 (accessed 9 August 2011).

Wampler, C., and K. Myhre (1990), 'Methodology for selecting PDC bits cuts drilling costs,' *Oil and Gas Journal*, **88** (3), 39–44.

Watts, R.L., S.A. Smith, J.A. Dirks et al. (1984), 'Photovoltaic product directory and buyer's guide,' Prepared for the US Department of Energy by Pacific Northwest Laboratory, April.

Wise, J.L., T. Roberts, A. Schen et al. (2004). 'Hard-rock drilling performance of advanced drag bits,' *Geothermal Energy – The Reliable Renewable, 28. Geothermal Resources Council Transactions*, 177–84.

Wiser, R., G. Barbose, and C. Peterman (2009), *Tracking the Sun: The Installed Cost of Photovoltaics in the U.S. from 1998 to 2007*, Berkeley, CA, USA: Environmental Energy Technologies Division, Lawrence Berkeley National Laboratory.

Witt, C.E., R.L. Mitchell, and G.D. Mooney (1993), 'Overview of the Photovoltaic Manufacturing Technology (PVMaT) project,' NREL/TP-411-5361, Golden, CO: National Renewable Energy Laboratory.

Witt, C.E., R.L. Mitchell, M. Symko-Davies et al. (2001), 'Ten years of manufacturing R&D in PVMaT—technical accomplishments, return on investment, and where we go next,' paper presented at the 28th IEEE PV Specialists Conference, Anchorage, AK, USA, January.

Yaws, C.L., K.Y. Li, and S.M. Chou (1986), 'Economics of polysilicon processes,' NASA Jet Propulsion Laboratory Proceedings of the Flat-Plate Solar Array Project Workshop on Low-Cost Polysilicon for Terrestrial Photovoltaic Solar-Cell Applications, Palo Alto, CA, pp. 79–121.

Zweibel, K. (2001), Thin-film partnership national research teams, NREL/CP-520-31064, Golden, CO: National Renewable Energy Laboratory.

Index